PROPER JIHAD

BY GREGORY HEARY

When determining whether a religion is violent or tolerant we have to examine the religion itself and what it originally taught. As difficult as it may be we cannot base our conclusions on what the people who claim to follow that religion do, nor can we base our conclusion on what the religion claims to teach today. This is because if its position has changed then it's a man-made religion and is subject to change again, according to social conditions. For example if someone were to place a grenade on my desk it would make me feel a little uncomfortable and I would ask why they put an explosive device on my desk. If they said *"It's not an explosive, it's a paperweight."* I would explain that it doesn't matter if you call it a paperweight or not, the fact remains that it was designed to explode and injure. Likewise with religion it might preach peace and tolerance today, but one has to examine what it originally taught in order to justify declaring it a peaceful or violent religion. Therefore it is important to explain the Islamic view towards violence, because sometimes people look at violence done in its name today and consider it as proof that Islam is intolerant and violent. Some mistakenly believe that Muslims force people to convert to their religion. Before I became a Muslim I didn't think I had ever personally met one, so nobody could have possibly forced me. Historically Islam was the first religion to declare people had free choice to choose their religion and teach religious tolerance. The Quran contains the first mandate for religious tolerance ever, in 2:256 it says:

لَآ إِكْرَاهَ فِى ٱلدِّينِ قَد تَّبَيَّنَ ٱلرُّشْـدُ مِنَ ٱلْغَىِّ فَمَن يَكْفُرْ بِٱلطَّـٰغُوتِ وَيُؤْمِنۢ بِٱللَّهِ فَقَـدِ ٱسْتَمْسَكَ بِٱلْعُرْوَةِ ٱلْوُثْقَىٰ لَا ٱنفِصَامَ لَهَا وَٱللَّهُ سَمِيعٌ عَلِيمٌ ۝

"Let there be no compulsion in religion: Truth stands out clear from Error: whoever rejects evil and believes in Allah hath grasped the most trustworthy hand-hold, that never breaks. And Allah heareth and knoweth all things."

In Islam there is "*NO COMPULSION IN RELIGION*". If someone wants to disbelieve in the truth and face the consequences in the afterlife then that is their decision. If the heart of a person has no belief then their actions are useless. Islam is a religion of voluntary submission, if you force someone to be a Muslim they will not be sincere and it will not count. The one who forces Islam upon someone will actually be considered sinful and oppressive, hated by Allah because Allah hates oppressors. No religion whether true or false may be forced upon someone. Even if that religion is popular or political in nature, such as democracy, republicanism, freedom, equality or Americanism. An entire chapter 6 verses long teaches this as well. It is chapter 109 of the Quran called Kafirun, which in English means "*The Disbelievers*" which is what the chapter is about. It details how Muhammad pbuh was instructed by Allah to deal with disbelievers:

قُلْ يَـٰٓأَيُّهَا ٱلْكَـٰفِرُونَ ۝ لَآ أَعْبُدُ مَا تَعْبُدُونَ ۝ وَلَآ أَنتُمْ عَـٰبِدُونَ مَآ أَعْبُدُ ۝ وَلَآ أَنَا۠ عَابِدٌ مَّا عَبَدتُّمْ ۝ وَلَآ أَنتُمْ عَـٰبِدُونَ مَآ أَعْبُدُ ۝ لَكُمْ دِينُكُمْ وَلِىَ دِينِ ۝

4

"Say : "O ye who reject Faith! I worship not that which ye worship, Nor will ye worship that which I worship. And I will not worship that which ye have been wont to worship, Nor will ye ever worship that which I worship.(as long as you aren't Muslim/don't believe in my faith of Islam) *To you be your religion, and to me mine."* "

No man could've come up with a better message to say to a disbeliever that combines permission for free choice and rejection for their disbelief. When this chapter was revealed it was in response to the pagans wanting the prophet Muhammad pbuh to compromise. The pagans would agree to be Muslim for one year if the Muslims would be pagan for one year, the pagans proposed to continue alternating religions each year so that everyone in the community would be united upon the same belief system yet both religions would be practiced bi-annually. Abu 'Abdullah ash-Shafi'i says that the verse *"To you your religion and to me mine"*, shows that the disbelievers are one people, because disbelief in all its many manifestations has one thing in common, its falsity. No other religion in the world teaches such a liberal message while at the same time reviling/rejecting disbelief. Islam teaches both tolerance and intolerance, Muslims respect people and their rights, but a Muslim will not show respect or support for disbelief in any form.

It is important not to take things out of context when discussing religions and violence. It is even more important to make sure that what is learned isn't out of context, particularly when one doesn't know anything about that particular religion. If you didn't know anything about

5

Christianity you might get the wrong impression if the only thing you heard was what the English New International Version of the Bible says in Luke 22:36 which attributes to Jesus pbuh:"36 *He said to them, "But now if you have a purse, take it, and also a bag; and if you don't have a sword, sell your cloak and buy one."* These verses taken out of context would make someone think Jesus pbuh was purely concerned with creating an army and that he taught weapons were more important than clothing. It would be unjust to take this verse of the bible and claim Christianity teaches militarization and aggression, yet this is similar to what the anti-Islamic Christians do to verses of the Quran when presenting their views. For instance there is a misconception that Muhammad pbuh came and spread Islam by the sword, yet the Arabic word for sword is not mentioned one time in the entire Quran, so something with this theory doesn't add up. When comparing religions many Christians who preach that Islam is violent say Christ pbuh came to spread peace on earth and goodwill towards men. To the ill-informed this would sound like Muhammad pbuh and Jesus pbuh had completely different messages and missions, but if we read the English New International Bible verse in Matthew 10:34 where Jesus pbuh is alleged to have said:"34 *"Do not suppose that I have come to bring peace to the earth. I did not come to bring peace, but a sword."* The bible paints a different picture than what Christians claim about Jesus pbuh. The bible contains the word sword 200 times, while the Quran does not contain the word sword once, even though Arabic has 12 different words for sword; these are undeniable literary facts. According to the bible the prophet Jesus pbuh in his own words *"did not come to bring peace, but a sword"*. This isn't the only instance in which Jesus pbuh is biblically portrayed as anti-peace either. In the English translation of the New International version of the Bible Luke 12:51 has Jesus pbuh saying, *"Do you think I came to bring peace on earth? No, I tell you, but division."* These are two times in two different gospels, in which the bible says Jesus pbuh explicitly stated he did not come to earth in order to bring peace, but a sword and division to which Jesus pbuh advises his companions to sell their cloak to buy a sword. Now does this mean Christianity is a violent religion and that Jesus pbuh was a genocidal mass murderer who forced people to believe in him or else be killed? Not necessarily. However if you were someone who didn't know anything about Christianity and I told you the bible said Jesus pbuh said these things then you would logically think that was the case. Especially if I said that Christ pbuh was an oppressive man who converted people by the sword, then played you a sample of the hymn played in churches that says, *"Onward Christian soldier, marching as to war..."* finally topping it off by quoting Luke 11:23 which says Jesus pbuh said: *"Whoever is not with me is against me, and whoever does not gather with me scatters."* The evidence is overwhelming, taken straight from the bible and church songs. How could anyone not view Christianity as a bloodthirsty murderous ideology when presented with the facts? Catholics even openly proclaim that they drink the blood of Jesus pbuh, they are literally thirsty for blood, even the blood of prophet Jesus pbuh. The bible clearly states that Jesus pbuh said he came with a sword to bring division and not peace where whoever is not with him is against him and those with him should sell their clothes to buy swords whilst those who don't gather with him "scatter". Sounds like quite

7

a violent intolerant guy and he was even known for outbursts of violence in his lifetime whipping people of a different religion than him in public places of worship such as a temple which they worshipped in. Oh and get this, Jesus pbuh is going to come back from paradise to kill somebody. Think about that, the prophet Jesus pbuh will leave paradise to kill. That's why gangsters say that Jesus pbuh is gangster. A Christian would say that selecting and presenting only these verses is completely unfair to the bible, Jesus (pbuh) and Christianity. Likewise it is unfair for people to quote a verse from the Quran out of context, deliberately trying to portray Islam as a violent religion spread by the sword and to tell lies about Muhammad pbuh to people who haven't learned about Islam before. If we were to use history as our only source of information and ignore religious teachings and books, then Christianity would be seen as much more violent than Islam. However just because Christianity has a bloody history doesn't necessarily mean the religion promotes violence. Likewise with Islam, some who are identified as Muslims commit acts of violence and even do it in the name of Islam, but their words and actions do not mean Islam teaches them to behave that way.

What is jihad? Jihad does not mean "holy war". In Arabic "holy war" is harb-u-muqadasah. During the crusades Christians called what they were doing a "holy war" and when Muslims did jihad in response they assumed it was the Islamic version of "holy war". Since the crusades were pretty much Christians killing anyone who wasn't Christian because they weren't Christian, by labeling jihad as "holy war" it gives people the misconception that jihad is the Muslim version of a crusade. It isn't. The word jihad has over 13 different meanings and only one of them has anything to do with war. That's means less than 8% of the meanings of Jihad involve warfare. Jihad means "to struggle" or "strive" against evil thoughts, evil actions, injustice and aggression against a person, family, society or nation. Now let's look at the "violent verses" so often cited against Islam to see what they say in context. The Quran says in 2:190-194 what means,

"Fight in the cause of Allah those who fight you, but do not transgress limits; for Allah loveth not transgressors. And slay them wherever ye catch them, and turn them out from where they have Turned you out; for tumult and oppression are worse than slaughter; but fight them not at the Sacred Masjid, unless they (first) fight you there; but if they fight you, slay them. Such is the reward of those who suppress faith. But if they cease, Allah is Oft-forgiving, Most Merciful. And fight them on until there is no more Tumult or oppression, and there prevail justice and faith in Allah; but if they cease, Let there be no hostility except to those who practise oppression. The prohibited month for the prohibited month,- and so for all things prohibited,- there is the law of equality. If then any one transgresses the prohibition against you, Transgress ye likewise against him. But fear Allah, and know that Allah is with those who restrain themselves."

While reading you might have noticed the 191st verse is translated as, "*And slay them wherever ye catch them*", this is the favorite verse of every anti-Muslim and they likely have it memorized, but they don't understand what it means.

First of all who is "*them*"? Does it refer to snakes, pigs or chickens? How could someone think it meant Christians? This is a quality of an extremist, they think every cry is directed against them. Just because there might be a "us vs. them" scenario that doesn't always mean you belong to either party. It's possible sometimes that you might not belong to "us" nor "them". Everything is not about you. Too many times the verse 2:191 is singled out and extracted out of context, distorted to give the impression that Islam orders Muslims to kill Christians and Jews because they practice a different religion. When these verses were revealed it was directly referencing the persecution by pagan Meccans oppressing Muslims just because they were Muslims. As you can see the word Jew or Christian is not mentioned within these verses anywhere. Specific instructions are given not to fight at the Kaba, sacred masjid, unless "they" fight you there. In the lifetime of Muhammad pbuh there were no Christian armies or Jewish armies planning to attack Muhammad pbuh at the Kaba when these verses were revealed. Although the pagans of Mecca were harassing Muslims there on a daily basis. Muslims had repeatedly asked Muhammad pbuh to let them fight back, but they were prevented from doing so because they did not have permission from Allah. Finally after so many years of patiently enduring persecution, Allah allowed the Muslims to defend themselves. Yet even still, Muslims were not given free reins to fight freely however they wished, there were conditions to be followed if one fought back. A Muslim could only fight those who fought Muslims. Muslims were not allowed to transgress the limits of warfare set by Allah, or else they would become of those whom

Allah doesn't love. The verse *"slay them where ye catch them"* is simple to understand, it has been forbidden to kill in the sacred masjid which is in Mecca since even before Muhammad pbuh was born, however this verse made it clear that it would not be sinful to kill an enemy of Islam who was persecuting Muslims if it was done inside the masjid at Mecca even though generally killing in that location is prohibited; this verse made an exception to the rule. The command was given to kill those specific people wherever they were found so their oppression would end. Being slain is the reward for those who suppress Islam, not for those who refuse to embrace Islam. Surely everyone would agree that if you are prevented from practicing your religion then you should be allowed to fight for your right to practice your religion. Even such wicked oppressors can still be forgiven by Allah. If they cease and desist then there is no justification to fight them, fighting them would then become a sinful crime. When the oppression of people exists Muslims will fight against it, when people are not oppressed a Muslim cannot fight. To say this allowance to fight is too much is to deny the right of self-defense. To say it's too little is to open the doors to oppression. Islam is a perfectly balanced religion, it is neither 100% aggressive nor 100% passive. It's not 50/50 either because 50/50 is not a perfect ratio regarding war and peace. Jihad and an Islamic State should not be the sole objective of a Muslim, it's important without a doubt, but the sole objective of a Muslim is the worship of Allah. The only reason an Islamic State is important is because it facilitates the worship of Allah. Thus an Islamic State is a sub-objective of Muslims to help fulfill the main objective. Jihad protects the religion of Allah and the rights of humans, as well as the Islamic State. Unfortunately not everyone who claims to be an Islamic State is an Islamic State. Many people call what they are doing jihad when it is not. Sometimes it can be confusing. Muslims must defend the faith and preserve the law of Allah. This is because Muhammad pbuh is the final messenger, so there will not be another update coming. Therefore if Muslims allow the faith to be extinguished then no one on earth will have access to the true religion, until the return of Jesus pbuh. If access to the true religion is lost then the people living at the time of the Second Coming would not even recognize Jesus pbuh if they saw him, but more importantly they wouldn't be able to worship God the way he desires. Therefore Muslims fighting against persecution and oppression is a matter of preserving the authority of God on earth. An Islamic State creates an Islamic environment which helps people worship God correctly by protecting humans from the sinful traps of our predator Satan. If Islam were to be eradicated chaos would rule over the earth. Chaos is practically the norm already in most countries, even those with populations who claim to have Muslim majorities, because few places are governed by Islamic laws. But those few places that are, have always flourished and experience order, stability, tolerance and happiness. Therefore whenever a place even resembling a potentially Islamic State exists the non-Muslim governments rush to crush it. Because if people witnessed a genuine Islamic State all other forms of government would be abandoned and many corrupt members of governments would have significant financial losses. This is why the pagan Meccans attacked the first Islamic State in Medinah because they didn't want that method of government to catch on. The forces of Satan do not want a country where the laws of God are the laws of the land. So Satan will use all the forces at his disposal to prevent a Islamic State as much as possible. Therefore Muslims must fight back and cannot turn the other cheek, because if we keep getting repeatedly slapped in the cheek eventually we will have a broken jaw and missing teeth; the ultimate result of continuous pacifism is death. Muslims are forbidden to kill themselves, thus if someone comes to kill you and you do not fight back then if you were to die that is basically suicide. It is explicitly explained in the Quran 4:29-30,

*"O ye who believe! Eat not up your property among yourselves in vanities: But let there be amongst you Traffic and trade by mutual good-will: **Nor kill (or destroy) yourselves**: for verily Allah hath been to you Most Merciful! **If any do that in rancour and injustice,- soon shall We cast them into the Fire**: And easy it is for Allah."*

The Quran is crystal clear, suicide is prohibited and the punishment for it is eternal hell. Remarkably the prophet Muhammad pbuh commented on Muslims who kill themselves, with many witnesses and different hadith testifying to the accuracy of the statement. Sahih Muslim hadith # 109 (a) says:

عَنْ أَبِي هُرَيْرَةَ قال قال رسول الله صلى الله عليه وسلم:

من قتل نفسه بحديدة فحديدته في يده يتوجأ بها في بطنه في نار جهنم خالدا مخلدا فيها أبدا ومن شرب سما فقتل نفسه فهو يتحساه في نار جهنم خالدا مخلدا فيها أبدا ومن تردى من جبل فقتل نفسه فهو يتردى في نار جهنم خالدا مخلدا فيها أبدا

Which in English means: "*It is narrated on the authority of Abu Huraira that the Messenger of Allah observed: "He who killed himself with steel (weapon) would be the eternal denizen of the Fire of Hell and he would have that weapon in his hand and would be thrusting that in his stomach for ever and ever, he who drank poison and killed himself would sip that in the Fire of Hell where he is doomed for ever and ever; and he who killed himself by falling from (the top of) a mountain would constantly fall in the Fire of Hell and would live there for ever and ever."*"

There is no dispute among Muslim scholars that suicide is forbidden in Islam and results in the person being eternally punished in hell.

Do people who claim to be Muslims kill themselves? Unfortunately some do, but suicide is still prohibited by Islam. I have listened to promoters of suicide bombings to learn how they justify them, which they dub "*martyrdom operations*" to see for myself on what basis they believe it's permissible, since they claim to be Islamic. Surprisingly they justify suicide bombing by quoting the bible. The bible contains a story about a mighty warrior called Samson who had supernatural strength, a woman seduced him and found out it was because he had long hair. Muslims don't believe power and strength comes from long hair, the Islamic belief is that there is no power nor strength except with Allah. Anyways the bible says that one night while this mighty Samson was asleep the woman plotted with his enemies and had his hair cut off, which caused him to lose his strength and become weak. He was then captured and taken prisoner by his enemies who saw it fit to take out his eyes. One day he was displayed as a prisoner of war at a lavish

14

dinner party for his enemies to triumph and gloat about his subjugation. Then the bible says he asked God to restore his strength to him so that he could knock over the pillars supporting the building thus killing himself and his enemies. Muslim extremists use this bible story to justify suicidal warfare as sanctioned by God. The real reason people claiming to be Muslims are blowing themselves up has absolutely nothing to do with Islam, it's because they read the bible and were misled by it. The Quran, as you have read, prohibits suicide explicitly. The Bible on the other hand has no such prohibition and can actually be interpreted as encouragement to wage suicidal warfare. For decades mutual destruction has been the fallback war plan of many Christian nations. In Christianity the first mention of suicide ever being sinful was not from the bible or any disciple or saint. The first instance of suicidal prohibition in the Christian tradition comes from Flavius Josephus, who was a Jew living after the time of Jesus pbuh. He wasn't even Christian! Today few dare to say that Christianity encourages suicide, but the idea of the crucifixion is essentially euthanasia. Most Christians believe God/son of God purposely sacrificed himself with fatal results for the good of the whole. This is the same thing extremists believe they are doing when they blow themselves up, one could say they got the idea from the Christians. Despite the stereotypes, Muslims did not start suicide bombing, it was prevalent among the Japanese in WWII and children blew themselves up in Vietnam trying to kill American soldiers. The Hindu Tamil Tigers used suicide bombings as a tactic since the 1980s CE in Sri Lanka, reportedly being the ones to have invented the suicide vest. In fact the first recorded

"Muslim" suicide bombing was done by the Iranian Shia during the Iraqi-Iranian war in the 1980s CE. Suicide bombing was first identified with Islam because of the Iranian Shia who used suicide bombings to kill Muslims. This makes Muslims the victims of suicide bombings long before they were ever considered perpetrators. Muslims were also the first people to have bombs dropped on them from planes, in 1911 CE when an Italian plane bombed Libya and in 1912 CE when Bulgarian planes bombed Turkey. Unfortunately that is how it is with most war tactics and Muslims, the Muslims tend to be the military guinea pig whom new weapons or tactics get tried on first. In actuality out of all the various war tactics used to hurt Muslims the suicide bombings done in the name of Islam hurt Muslims the most, because not only does it tarnish the image of Islam, but it causes severe retaliation in turn that kills more Muslims than the number of "enemies" who died in the suicide blast. A suicide bomber is like a bee who attacks somebody near a bee hive, the bee dies due to the attack and then the stung person seeking revenge destroys the whole bee colony killing all the bees because some stupid bee thought it was nobly sacrificing itself "for the greater good". Except the suicide bomber is frequently worse than the stupid suicidal bee because the suicidal bee never kills other bees in its attack. Now some may think, well if the "Muslim nest" is attacked then maybe for self-defense suicide bombing is allowed? Nope because suicide is not self-defense it's self-destruction. A Muslim is fundamentally a slave of God and as such they do not have the legal right to choose to sacrifice themselves for the sake of others to the extent that their sacrifice is explicit self-mutilating suicide.

Suicide is self-mutilation. Plus on the face of it if you are killing yourself to attack then you will not leave that battlefield victorious. A suicide attack has a risk of losing 100% of your life and a tiny % chance of possibly inflicting some damage on an enemy target. So the ratio of risk to reward is not worth it statistically speaking. If every soldier was a suicide attacker then there would be no need for the enemy to even fight, so that in itself reveals the lunacy of the tactic. Therefore to ever achieve victory, another tactic must be utilized so why not just use the winning tactic from the very beginning and get good at that instead of wasting time, energy and manpower killing your own troops? The reason for defense is to stop Muslim bloodshed, so if the attack requires Muslims to die then that is not an attack worth making. The life of 1 Muslim, nay 1 drop of Muslim blood, is more valuable than the death of all of the disbeliever's army. Likewise bees have no other military capabilities but humans have many ways to wage warfare without killing themself. To kill oneself is not jihad, jihad is to struggle, a suicide attack is the easy way out. A suicide attack is saying, *"I can't beat em so I'll take em with me."* In reality Satan can't beat the believer so he takes them out via getting them to destroy themselves with sins like suicide. Suicide attacks make Satan happy and give him company in the hellfire. The other justification Khawwarij extremists use is that because Muslims can't easily defend themselves from drones and the advanced weaponry of their enemies, desperate times call for desperate tactics. This is a blatant fallacy. When Moses pbuh was trapped between the sea and the might of Pharaoh's army he didn't send out suicide attackers to destroy the enemy mutually. Allah provided victory in an unexpected way using the sea as an escape route to safety as well as a weapon against the enemy. Who would ever think that water would destroy a military superpower? A similar situation occurred with David pbuh when he killed Goliath. David pbuh used inferior weapons yet with Allah on his side he was victorious. Likewise a lightning bolt, tornado, hurricane, earthquake, tsunami or volcano can destroy all the advanced weapons of modern man. What could a nation at war with Muslims do if an insect species attacked them? (as will be the case with the nations of Juj and Majooj) Or if Allah caused it to continuously snow on the enemies of Muslims for 5 months in a row so that their entire nation was covered in snow several thousand feet deep? Snow or rain alone can completely destroy any country in the world today if Allah just sends it down upon them continuously for an extended period. If Allah wills victory for Muslims he could send Angels to help them like at the battle of Badr, which the disbelievers saw with their own eyes. Today the only reason angels don't help the Muslims with their problems is because the angels are too busy recording our sins. As soon as the angels are done writing down our sins and have spare time then God has decreed that they help the Muslims defeat their enemies and establish Shariah. We've just been keeping the angels too busy writing our sins and they have to finish doing that before following God's command for them to help us. The same applies for the rest of the environment/creatures that God created to witness our sins so as to bear witness against us on the Day of Judgement. Today Muslims are sinners and never in all of history have sinners been winners. Military victories go to Muslims who have virtues not vices. Based on the teachings of the Prophet Muhammad pbuh, if you kill yourself with a steel weapon or by intentionally falling off a mountain you'd be in hell forever. So let's say your enemy is behind you and the only way to attack them is to thrust your steel sword through your own body in order to have it harm your enemy, thereby killing yourself. That would be a suicide attack. As would jumping off a mountain and dying from the impact of falling on an enemy's head. Such tactics were never used by the early generations of Muslims. When a person blows themselves up they are the first ones to die, then their deeds stop because they are dead, there is no guarantee any "enemy" will even be hurt and most of the time it is innocent civilians who get killed by suicide bombers. The effects of which only increases the oppression the extremists claimed they were trying to fight. These suicide attackers are just giving the enemies of Islam more ammunition to use as a basis for further persecution. Personally I believe the Khawwarij extremists subconsciously want to increase the persecution, surely they cannot be so blind that they fail to see the results these suicide attacks have. Suicide attackers are not only disobeying Islam but they are harming Islam and Muslims, whether they realize it or not. May Allah protect us from destroying ourselves in any manner. I'm sure you might have heard it many times before that Muhammad pbuh taught Muslims how to wage war. It's true he taught Muslims how to battle, but opponents of Islam will never tell you what the rules which he taught Muslims to follow while waging war were. Concerning warfare, since the 600s CE Islamic Sharia legislated the following rules of engagement for Muslims when fighting Jihad in a military confrontation:

- It is prohibited for a Muslim warrior to kill women, the elderly, the disabled or children. (unless they are part of the enemy's army and attacking you, if they are innocent civilians it is a major crime and sin to kill them)

- Muslim warriors cannot kill grazing animals, unless it's out of necessity for food and then they must be paid for.

- Muslim warriors are forbidden to rape or plunder.

- Muslims cannot mutilate a dead body of any creature.

- Non-military facilities cannot be used by Muslim warriors without permission from the natives, if granted the natives must be financially recompensed.

- Only active combatants may be combated.

- It is forbidden to kill a wounded incapacitated enemy.

- When an enemy surrenders it is illegal to harm them.

- Prisoners of war must be well fed, provided with good clothes, proper medical care and have loose shackles.

- Nobody can ever force someone to become a Muslim.

Now which country today also follows these rules of warfare? Certainly not America, the American military and the UN "peacekeepers" break these rules routinely. They break these rules so much that they aren't even rules in their warfare rulebooks. Thus in their eyes they aren't breaking these rules because they don't believe in the rules for war.

As a result their soldiers suffer from PTSD formerly known as Operation Exhaustion, Battle Fatigue and Shellshock; God knows what they'll call it in the future. The reason such soldiers suffer from mental and emotional problems is because they don't follow the rules of jihad. As with anything one does which God has made illegal, there are negative side effects. So when people fight in an unislamic manner they will have negative side effects. Whereas Islamic Muslims who wage jihad have never in all of history suffered from any mental, emotional or social disorders as a result of combat, yet their enemies always have. When you unbiasedly compare Islamic warfare to how non-Muslims wage war, non-Muslim warriors are actually the real evildoers. Although they cover it up by saying "*all is fair in love and war*". Islam says there are rules to follow even in love and war. When you love a person there are conditions on how you can treat them, if they don't love you back then you have no right to rape them, just because you love them it doesn't give you the right to abuse or oppress them. Likewise in warfare you cannot fight by all means necessary; even if your enemy does. No atomic bombs or civilian casualties are allowed. Only combatants can be harmed if they are currently engaging. If the enemy surrenders, then legally you have no right to harm them and they become prisoners who must be treated respectfully with compassion. Oppression must not occur. "*Collateral Damage*" is unacceptable, sinful and impermissible for Muslims. In Islam anyone who intentionally kills an innocent person is treated as a murderer and punished for it. What does the bible say about killing noncombatants such as women and children?

Numbers 31:7-18 "*7 They fought against Midian, as the Lord commanded Moses, and killed every man.* *8* Among their victims were Evi, Rekem, Zur, Hur and Reba – the five kings of Midian. They also killed Balaam son of Beor with the sword.*9 The Israelites captured the Midianite women and children and took all the Midianite herds, flocks and goods as plunder.10 They burned all the towns where the Midianites had settled, as well as all their camps.11 They took all the plunder and spoils, including the people and animals, 12 and brought the captives, spoils and plunder to Moses* and Eleazar the priest and the Israelite assembly at their camp on the plains of Moab, by the Jordan across from Jericho.*13* Moses, Eleazar the priest and all the leaders of the community went to meet them outside the camp.*14* **Moses was angry with the officers of the army** – the commanders of thousands and commanders of hundreds – who returned from the battle.*15* **"Have you allowed all the women to live?" he asked them.***16* "They were the ones who followed Balaam's advice and enticed the Israelites to be unfaithful to the Lord in the Peor incident, so that a plague struck the Lord's people.*17 Now kill all the boys. And kill every woman who has slept with a man,18 but save for yourselves every girl who has never slept with a man.*"

The english translation of the New International Version of the Bible states only the virgin women were spared from the slaughter. How did the biblical soldiers determine which women were virgins? The key word revealing the answer to that question is "were". However that's just 1 example, it would truly be unjust to cite just 1 biblical example of warfare. Besides that is just an example, it's not like that's the standard rules for war which the bible teaches is it? It is. As explained in Deuteronomy 20:1-20,

When you go to war against your enemies and see horses and chariots and an army greater than yours, do not be afraid of them, because the Lord your God, who brought you up out of Egypt, will be with you. ² When you are about to go into battle, the priest shall come forward and address the army. ³ He shall say: "Hear, Israel: Today you are going into battle against your enemies. Do not be fainthearted or afraid; do not panic or be terrified by them. ⁴ For the Lord your God is the one who goes with you to fight for you against your enemies to give you victory." ⁵ The officers shall say to the army: "Has anyone built a new house and not yet begun to live in it? Let him go home, or he may die in battle and someone else may begin to live in it. ⁶ Has anyone planted a vineyard and not begun to enjoy it? Let him go home, or he may die in battle and someone else enjoy it. ⁷ Has anyone become pledged to a woman and not married her? Let him go home, or he may die in battle and someone else marry her." ⁸ Then the officers shall add, "Is anyone afraid or fainthearted? Let him go home so that his fellow soldiers will not become disheartened too." ⁹ When the officers have finished speaking to the army, they shall appoint commanders over it. ¹⁰ When you march up to attack a city, make its people an offer of peace. ¹¹ If they accept and open their gates, all the people in it shall be subject to forced labor and shall work for you. ¹² If they refuse to make peace and they engage you in battle, lay siege to that city. ¹³ When the **Lord** your God delivers it into your hand, put to the sword all the men in it. ¹⁴ As for the women, the children, the livestock and everything else in the city, you may take these as plunder for yourselves. And you may use the plunder the Lord your God gives you from your enemies. ¹⁵ This is how you are to treat all the cities that are at a distance from you and do not belong to the nations nearby. ¹⁶ However, in the cities of the nations the **Lord** your God is giving you as an inheritance, do not leave alive anything that breathes. ¹⁷ Completely destroy them — the Hittites,

Amorites, Canaanites, Perizzites, Hivites and Jebusites — as the Lord your God has commanded you. [18] Otherwise, they will teach you to follow all the detestable things they do in worshiping their gods, and you will sin against the Lord your God.

(Remember this rule in Deuteronomy 20:16-17 when Jews, Christians and Zionists say God has given/promised the Jews Palestine and other areas of the middle east as an Inheritance. That's why Israel "completely destroys" the things living in those lands today not leaving anything alive that breathes. Its due to this biblical ruling to kill every non-Jew in the "Promised lands". The "promised lands" came with a promise of gentile genocide, and that's a Jewish genocide of gentiles not to be done gently. The religious genocide by Israel is real and violent by design. Their "peace process" is about how to get more pieces of land. They don't want peace in the holy land they want a piece of it.)

[19] When you lay siege to a city for a long time, fighting against it to capture it, do not destroy its trees by putting an ax to them, because you can eat their fruit. Do not cut them down. Are the trees people, that you should besiege them? [20] However, you may cut down trees that you know are not fruit trees and use them to build siege works until the city at war with you falls." (You see fruit trees have more rights than non-Jewish civilians do according to the biblical text.)

Since we quoted the rules for biblical warfare as allegedly waged by Moses pbuh, with 1 of the many biblical examples that indicate he waged ruthless warfare, lets see examples of biblical warfare allegedly waged by Joshua and David pbut.

24

Joshua 10:29-40, "*Then Joshua and all the Israelites traveled from Makkedah to Libnah and attacked that city. [30] The Lord allowed the Israelites to defeat that city and its king. They killed everyone in the city. No one was left alive. And they did the same thing to that king as they had done to the king of Jericho.[31] Then Joshua and all the Israelites left Libnah and went to Lachish. Joshua and his army camped around that city and attacked it. [32] The Lord allowed them to defeat the city of Lachish. They defeated it on the second day. The Israelites killed everyone in the city, just as they had done in Libnah. [33] King Horam of Gezer came to help Lachish, but Joshua also defeated him and his army. No one was left alive.[34] Then Joshua and all the Israelites traveled from Lachish to Eglon. They camped around Eglon and attacked it. [35] That day they captured the city and killed everyone in the city. This was the same thing they had done to Lachish.[36] Then Joshua and all the Israelites traveled from Eglon to Hebron and attacked it. [37] They captured the city and all the small towns near Hebron. The Israelites killed everyone in the city, just as they did to Eglon. No one was left alive there. They destroyed the city and killed all the people in it as an offering to the Lord.[38] Then Joshua and all the Israelites went back to Debir and attacked it. [39] They captured the city, its king, and all the towns near Debir. They killed everyone in the city, just as they had done to Libnah and its king. No one was left alive there. They destroyed the city and killed all the people in it as an offering to the Lord. [40] So Joshua defeated all the kings of the cities of the hill country, the Negev, the western foothills, and the eastern foothills. The **Lord**, the God of Israel, had told Joshua to kill all the people, so Joshua did not leave anyone alive in those places.*"

1 Samuel 18:22-27, "*Then Saul ordered his attendants: "Speak to David privately and say, 'Look, the king likes you, and his attendants all love you; now become his son-in-law.'"[23] They*

repeated these words to David. But David said, "Do you think it is a small matter to become the king's son-in-law? I'm only a poor man and little known." [24] *When Saul's servants told him what David had said,* [25] *Saul replied, "Say to David, 'The king wants no other price for the bride than a hundred Philistine foreskins, to take revenge on his enemies.'" Saul's plan was to have David fall by the hands of the Philistines.* [26] *When the attendants told David these things, he was pleased to become the king's son-in-law. So before the allotted time elapsed,* [27] *David took his men with him and went out and killed two hundred Philistines and brought back their foreskins. They counted out the full number to the king so that David might become the king's son-in-law. Then Saul gave him his daughter Michal in marriage.*

When we compare jihad with the bloody biblical warfare, we see that Islamic warfare is more moral and safer for civilians. In fact militarily speaking the conquests of Muhammad pbuh were the most peaceful in recorded history. Anti-Muslims claim the "war period" began the first day Muhammad pbuh went to Medinah and lasted til his death 10 years later. This is a false claim that he waged war every day for 10 years, but we will pretend it were true for the sake of their argument. In this "decade of war" the conquests of Muhammad pbuh amounted to over 3 million square kilometers. This averages out to mean conquering 900 kilometers per day for 10 years. In these 10 years there were 63 total battles, most of which he didn't participate in personally, which resulted in 1,287 total casualties. The number of disbelievers killed on the battlefield during these 10 years are 1,022 total whilst there were 265 Muslim warriors who died on the battlefield. Mathematically that means on average about 16 disbelievers and about 4

Muslims were killed per battle, even though in many of the battles(including the most important ones) the Muslims were outnumbered by a wide margin, with inferior weaponry and equipment as well. If you account for the numbers of how many fought in the battles the total amount of fighters killed on both sides equals less than 1%. This was during the 7th century CE. Whereas today frequently more than 100% of the number of fighters die in wars because so many non-combatants die. Some modern "civilized wars" even have 300% or 500% casualty rates because the wars result in so many more civilians dying than there are soldiers fighting. Ask anyone in the military profession if it is possible to conquer 3 million square kilometers whilst killing less than 1% of the hostile enemy forces who are zealously fighting you for religious reasons? They will say the only way this could possibly happen is if they were clear good guys and their enemies were clear bad guys and that they didn't oppress people or give any possible reason whatsoever for those whom they conquered to be dissatisfied or displeased. Those conquered would actually have to be happy they were conquered or else such a campaign would be unfeasible. Yet even then one would need a miraculous army to be able to conquer such a vast amount of territory in such a short period of time with so few deaths on both sides, unless God was literally on their side and the army was just, good, righteous and strictly tried their utmost to keep bloodshed to the absolute minimum. An army of bad guys could never do it, good guys couldn't even do it and it would be extremely difficult if not impossible for great guys to do it. They say you would need somebody like Moses pbuh or Jesus pbuh to be the leader of such an army to get such results. Yet who was this leader? Muhammad pbuh.

Does Islam teach Muslims to be violent? Islam teaches the right of self-defense and self-defense is a type of violence. This world has evil people who oppress others by force. Islam is realistic and provides the solution for oppression. When people are starving and dying in a country rich with many natural resources, Muslims do not just give food aid and hope it solves the crises. Muslims ask *"What/Who is causing the problem?"* and we strive to find a permanent solution; we cure the problematic symptoms by curing the disease. Sometimes violence is the only thing that can stop the plagues of unjust violence and oppression. Muslims cannot just sit and supplicate to Allah hoping the problems go away, Allah has legislated jihad as a tool for humans to stop oppression throughout the universe. Islam teaches Muslims to stop injustice. So Jihad is not only for self-defense but also to defend the rights of others as well as the rights of God. While the rights of God are more important than the rights of people. However Islam places severe restrictions on violence so that a Muslim warrior does not transgress the bounds set by the Creator or wage immoral warfare. The Quran explains the wisdom of Allah behind these restrictions in 60:7-9, "*It may be that Allah will grant love (and friendship) between you and those whom ye (now) hold as enemies. For Allah has power (over all things); And Allah is Oft-Forgiving, Most Merciful. Allah forbids you not, with regard to those who fight you not for (your) Faith nor drive you out of your homes, from dealing kindly and justly with them: for Allah loveth those who are just. Allah only forbids you, with regard to those who fight you for (your) Faith, and drive you out of*

28

your homes, and support (others) in driving you out, from turning to them (for friendship and protection). It is such as turn to them (in these circumstances), that do wrong."

It is crucial to clarify that not just any Muslim can declare jihad. Only a Muslim leader of an Islamic state is qualified to declare jihad, after being advised by Muslim scholars on the conditions and rulings of jihad. A Muslim without authority has no right to declare a jihad. An individual Muslim could still wage jihad in special circumstances, such as if Muslims or innocents were physically attacked then the Muslim would be obligated to do jihad and defend them; but that is not the same as declaring jihad. The manner in which jihad is declared is legislated by Islam. One cannot just "go to war" even if it is waged appropriately and religiously ordained. Jihad has to be declared in a specific fashion. A Muslim envoy or ambassador on behalf of the Muslim ruler must be sent to the leader(s) of those whom qualify for jihad to be waged against. The envoy then informs them of the three options they have:

1. <u>Embrace Islam and become Muslims</u>, which would entail stopping whatever oppression and wrongdoing they would be engaged in. If they embrace Islam they're invited to live in the Muslim lands and if they migrate they'll have all the privileges and responsibilities of the first Muslims along with a share in all spoils of war. If they become Muslim but don't migrate then they will not be entitled to any share in the spoils of war, unless they participate in the actual fighting of war.

2. <u>Don't Embrace Islam, but pay the jizya</u> and become a vassal of the Islamic state, which would entail stopping the corruption within that society as well as allowing Muslims to fully practice Islam 100% in those lands without harassment and to propagate it. (Jizya is a small fee paid by non-Muslims which entitles them to practice their religion in a Islamic state in private without harassment from Muslims. Whoever can't afford it doesn't pay anything, just like with zakat. Zakat is more expensive, it's actually more expensive to be a Muslim in an Islamic state than to be non-Muslim.)

3. <u>Don't accept options 1 or 2 and we will fight</u> until the reasons for waging jihad no longer exist.

That's what Islam teaches. But has anyone actually ever followed the rules of Islam when waging jihad? An incident that occurred during the Khilafah of Umar Ibn Abdul Aziz who ruled from 717 CE -720 CE demonstrates how a proper Khalifah of an Islamic State wages jihad. When he was in power the Muslim nation extended from China to the Atlantic ocean and the Muslim army led by Qutaybah ibn Muslim had conquered the city of Samarkand. The way they did this was through a surprise attack from behind the mountains so that the disbelievers were unable to see them coming and couldn't properly prepare a defense. After they took the city the Muslims left the population entirely unharmed and no damage was done to anything in the city, the Muslims actually helped the disbelievers go about their daily routine. After seeing how the Muslims treated them with justice, the disbelievers sent a person to the Khalifah Umar ibn Abdul Aziz to complain about how the city was conquered. The man argued with the Khalifah that he had learned when Muslims wage Jihad they have three options Islam requires Muslims to give disbelievers before conquering them: First invite them to Islam. Second ask them to pay tribute(jizya) and in the event of their refusal of both the choices, then you give them the choice of war. Umar said that was true and every country has the right to choose among the three. The man said: "*Is it in your custom to start the assault by surprise?*" Umar ibn Abdul Aziz said: "*It is not our custom to do so and Allah Almighty has ordered us not to do so, and our Prophet forbade us from being unjust.*" The man said in response, "*Qutaybah ibn Muslim did not do what you say and his army attacked us by surprise.*" When the Khalifah confirmed this was true he sent instructions to the new Muslim governor of Samarkand, Sulayman bin Abi Sarri, to appoint a Muslim judge to give the correct Shariah judgment ruling on the case and for the governor to do what the judge said. The case was brought to the Shariah court to decide between the priests of Samarkand representing the disbelievers and the Muslim general Qutaybah ibn Muslim standing next to them representing the Muslims. A priest made his case that Qutaybah ibn Muslim entered Samarkand without warning them. He didn't give them warning nor the options of invitation to Islam, or payment of tribute, or war. Instead, he attacked without warning. The judge, Jumay'a bin Hadir, turned to the commander Qutaybah ibn Muslim and asked him: "*What do you say about this complaint?*" Qutaybah said: "*War is a trick. This country is a great obstacle to us and all those who were like it did not pay tribute and did not want to enter*

Islam, and had we fought them, they would have killed more of us than we would have killed from them." He ardently continued, *"And by the help of Allah and surprise, we defended Muslims from great harm and the history is witness to what I say. And all the countries beyond them became easy to conquer. Yes, we surprised them but rescued them and let them know about Islam."* The judge asked: *"Qutaybah! Did you invite them to Islam, tribute or war?"* Qutaybah replied, *"no, we surprised them for what I told you before."* The judge declared: *"Qutaybah, you confessed and by this the court's duty ends. Qutaybah, Allah supports this nation only by religion and by avoiding treachery, and setting up justice. We were out of our homes for jihad for Allah's sake. We didn't go out to conquer lands and occupy countries unjustly."* Then the judge issued his ruling: *"I rule that all armies of Muslims in this country should get out of this country and give it back to its people and give them the opportunity to prepare for war, and then make them choose between Islam, tribute or war. If they choose war then we will fight. Muslims will get out of Samarkand without anything just as they entered and deliver the city to its people, and that is the application of the law of Almighty Allah and the Sunnah of our Prophet Muhammad peace be upon him".*
After this, Muslims started to leave the city. The judge came out in front of the priests. The priests did not believe what they were seeing. The people of Samarkand kept watching Muslims leave until they saw that all of them had left their city. Afterwhich a priest said, *"What they did proves that their religion is the right. I witness that there is no god but Allah and that Muhammad is His Messenger."* Then all the priests declared there is no god but Allah and that Muhammad is His Messenger (pbuh), thereby becoming Muslims along with many of the inhabitants of the city. Afterward they

asked the Muslims to come back, teach them Islam and voluntarily chose to be part of the Muslim Nation. This is how an Islamic State acts and wages jihad. You can check the history books, nowhere else in all of history has a country who conquered another country unfairly then appointed a judge from their own country to rule in a case due to a complaint from one of the conquered in which the judge ruled against his own country and the conquering country voluntarily left as a result. The Islamic Khilafah is the only country in the world to have ever acted in such a manner. Check the history books! Muslims are the only ones who have ever acted with such justice when they followed Islam correctly and waged jihad correctly. So those today who say they don't want an Islamic Khilafah or condemn jihad are either extremely ignorant of Islam or completely crazy. Even if you are non-Muslim why wouldn't you want a country to act like how Islam teaches and how the Islamic Khilafah acted? Umar Ibn Adul Aziz ruled generations after the prophet Muhammad pbuh and his life should be studied by every Muslim and non-Muslim so they know how a Muslim ruler and Khilafah should operate; especially in regard to how he dealt with Khawarij extremists. Umar ibn Abdul Aziz became the leader of the Muslims in September 717 CE and died in office during February 720 CE. I encourage you to read his story and I guarantee that even if you hate Islam and Muslims you will wish your country had a leader like Umar ibn Abdul Aziz, you will likely cry after learning how just and humble he was. As leader of the Islamic State he did the following:

- He abolished taxes, and instead requested all the governors to encourage the citizens to practice agriculture.

- He dictated if anybody had a piece of land which he/she did not cultivate for three years, the land was to be taken from them and given to somebody else who would cultivate it.

- State officials were banned from being involved with businesses. Government was to have absolutely zero ties to businesses.

- Unpaid labour was made illegal. All who labored(including slaves) were paid before the sweat on their back had dried, in accordance with how Muhammad pbuh taught laborers should be paid.

- Pasture lands and game reserves (which had been previously unislamically reserved for families of high ranks) were evenly distributed among the poor for the purpose of cultivation.

- He urged all officials to listen to the complaints of their citizens. Furthermore Umar ibn Abdul Aziz used to announce if any subject had seen an officer mistreating the people, the officer should be reported to the leader and him and that subject would be given a reward ranging from 100 – 300 dirhams. Meaning if you saw a government employee do something wrong and mistreat a single person, it was your duty to report it to their superior and the head of state. When you did then you would get paid 100-300 silver coins and the guilty official would be reprimanded and punished for it.

- He outlawed handcuffs because they prevented criminals from performing their obligatory prayers and/or voluntary. So apprehended criminals were not handcuffed because to hinder them from praying was made a crime in and of itself. For any government employee to put handcuffs on a criminal would result in them being punished for crime themselves. What's called "Police brutality" today did not occur because handcuffs themselves were considered a form of criminal police brutality, so unjust police violence was nonexistent. Criminals had the right to not be handcuffed in Umar's state.

There are 4 categories of people concerning Jihad.

1. **Dhimmi**: This is a non-Muslim citizen living in a Islamic State. A dhimmi's life and property is protected, and it is forbidden for Muslims to commit any injustice to them. Muslims are permitted to treat them kindly, especially in worldly matters, without honoring their religions or honoring them as we honor Muslims. Muslims and Dhimmis are not equal spiritually nor legally, so they are not treated equally. Yet Dhimmis still have rights and cannot be oppressed despite being a second-class citizen due to their choice to believe/practice a false religion.

2. **Muahid**: This is a non-Muslim who has a treaty with Muslims. Ex. A non-Muslim from an unislamic nation which has signed a peace treaty and isn't at war with Muslims. The Muahid enjoys extra rights which the non-Muslims in non-Muslim states who don't have a peace treaty with Muslims do not enjoy. Any disbeliever who lives in a nation which is at peace with Muslims is given extra rights due to being part of a peaceful nation. Islam rewards non-Muslim citizens of peaceful nations. If a non-Muslim is a citizen and resident of a nation that is at war with Muslims they are not considered a Muahid. So therefore any unislamic or non-Muslim government which is fighting Muslims/Islam deprives it's citizens of rights which they would enjoy if their government were at peace with Muslims/Islam instead of being at war. Basically non-Muslims, through various means, can and do actively choose which of these 4 categories they fall into and each category is afforded different rights.

3. **Mustamin**: This is someone who has been given specific protection by Muslims. Traditionally an example of a Mustamin would be a businessman given a visa pass to enter a Muslim country to do business. (True Islamic countries don't have tourism industries or allow tourists to visit.) In an Islamic state that truly implements the Shari'ah, rulings pertaining to a mustamin are similar to those for a dhimmi, for the most part. They will be obligated to pay the jizyah if they remain in the Islamic state for a year. Payment of the jizyah is a condition to

be classified as a dhimmi. A Mustamin could pay Jizyah depending on the duration of their stay in an Islamic country but while Jizyah is a condition of being a dhimmi it is not the only condition. So a Mustamin could pay Jizya and still not be a dhimmi. Thus the legal rights of a Mustamin would be different, because they would be a foreigner and not a citizen.

Those three types of people are off-limits in Islamic warfare. A Muslim who harms any of those types of people would be considered sinful unless there were a just reason, such as self-defense or if it were accidental. Simply being a non-Muslim in a Muslim state or Islamic state doesn't make one a Dhimmi, but it doesn't mean one can't be safe either. Likewise for any non-Muslim to be on the "safe list" they don't have to live in a non-Muslim country or become a Muslim. Sadly though sometimes unislamic non-Muslim governments can put their non-Muslim citizens into danger by waging war against Muslims. Although fortunately those first three types of non-Muslim people cannot have any military war declared upon or waged against them. If you are a non-Muslim in one of the 3 previously mentioned categories then, barring any illegal activities you may have committed such as murder/rape or something, you are safe from any and all physical harm and no Muslim could ever legally harm you from an Islamic perspective. To harm the above 3 types is a sin.

4. **Muharab**: This is someone who declares war on you. The non-Muslim who declares war on Islam and/or Muslims who actively fights against them must be fought against, until the Muharab is dead or stops

fighting. An example of this type would be a Crusader.
Traditionally non-Muslim citizens of Muharab states
were considered Muharabs. However there are
different categories of Muharabs. For example a baby, a
cripple or a woman, a child, an elderly person, a
priest/rabbi and an insane person are all considered to
belong to a special class of Muharab different from the
fighting class of Muharab. The non-Muslim citizens of
a Muharab state who are not actively assisting the
Muharab state in it's efforts of harb(war) is different
than those who are. For example a non-Muslim
Muharab citizen who opposes the war effort, or spies on
behalf of the Muslims or hinders the war effort of the
Muharab state is a different type than those who join
the Muharab military, support it or is indifferent to it.
So there are about 3 types of non-Muslim Muharab
combat capable non-fighter citizens. 1. Those who
combat/oppose the Muharab state either via peaceful
or violent methods. 2. Those who are neither
supporting nor combating the war which the Muharab
nation is waging against Muslims. This category is very
rare, usually it's like a prisoner or jungle nomad who
has no clue what is going on in the world and no ability
to do anything regardless of their opinion. Thus this
indifference is due sheerly to circumstances not by
attitude. In Islam political neutrality regarding war
typically can only be due to extreme ignorance of a war,
if you know there is a war then there is no neutrality
unless the person entirely secludes themself from being
able to effect the conflict in any way whatsoever. So the
neutral non-Muslim citizen of a Muharab state is

extremely rare, especially in modern times with mass taxation. 3. Those who are supporting the war against Islam/Muslims which is waged by the Muharabs. This support can be done in various types of methods, politically, financially, medically, emotionally, religiously. Each of these 3 categories has different rulings regarding their rights during a Jihad. Some are considered combatants and some such as those in category 1 are treated as non-combatants entitled to rewards from the Muslims. However it does get tricky when labeling people because sometimes people can belong to more than one category of the categories of Muharabs. For example someone could be making weapons for the Muharab military to use on Muslims yet then they go out protesting against the war. So such a person would be in both category 1 and 3. Muslim scholars have different opinions over the treatment of the 3 different categories, particularly because within the 3 general categories there are categories. Such as category #1 being anti-war has many different levels, some of which could be considered treason by the Muharab state while others would be considered legal or even praisworthy by the Muharab state. Similarly there are different categories of #3, such as the wife of a Muharab soldier having intimacy with her husband. Technically that's "supporting the military which is fighting Muslims" but on the other hand is that enough support to give her the legal label of assisting the enemy of Muslims to the extent where she becomes a valid military target? It all depends on many variables including all the other stuff she does or doesn't do. For

example if she was having sex with her soldier husband
in order to delay him from reaching the battlefield or
hoping he gets an STD or something then that's
different. Therefore the categories of the citizen of a
Muharab state can be very vast regarding their status as
military targets. Generally a Muharab is to be fought
until dead or until they are no longer a Muharab,
however the 3 categories of non-military Muharabs are
not as easy to define and require a vast amount of
knowledge and wisdom particularly since it involves
matters of life, death and bloodshed. So if one is a non-
Muslim citizen of a Muharab state and you aren't a part
of the Muharab military different rulings still apply that
could endanger your safety or life depending on how
the person reacts to the war the Muharab state is
involved in. It is possible to be a citizen and resident of
a Muharab state and still be entitled to safety from
Muslims waging Jihad. However it is also possible to
be a non-soldier citizen and resident of a Muharab state
and be at risk of bodily harm in the course of the war.
The safest thing for a non-Muslim is to not be in/of a
Muharab state. As a result of the war of a Muharab
state they are stripping their citizens of rights and
putting them nearer to and sometimes in real danger
from violent Jihad. When governments go to war the
status and safety of their citizens changes as a result of
that war. After the war the safety/status of their
citizens may improve or decrease but war is an affair
wherein the levels of rights and safety which citizens
enjoy fluctuates until the end of the war and in
accordance to their actions during that war. If a

disbeliever never wants to be in danger from any Jihad then they just can avoid being a Muharab or supporting the Muharab war efforts in any way, and/or actively oppose the Muharabs and their war efforts without giving any support. Unfortunately sometimes accidents in war happen and civilians can get caught in the crossfire or mislabeled. So generally Muslims fight Muharabs, but it's complicated. However Islam teaches that Allah commands certain types of violence is to be meted out to certain types of Muharabs. Don't be a Muharab, don't even be friends with a Muharab. Being a friend of a Muharab can even endanger your life or afterlife. War is a polarizing thing that changes human relationships, sometimes for the better and sometimes for the worse. But war always changes the relationships between citizens of different nations sometimes with bloody consequences. Hence Islam teaches Allah teaches humans how to treat other people during war so they treat them as God wants. If you always treat people how God wants then everything will turn out good for you in the end, after the test is over. Muharabs may not like Muslims to treat them as God says to, but Muslims, being slaves of God, treat them as ordered.

Whoever gives protection to a non-Muslim Dhimmi, Muahid or Mustamin then kills them, deserves the curse of Allah, the curse of the angels and the curse of all mankind; on the day of judgment nothing will be accepted from them. This is why jihad is risky business which if someone is doing they must be 100% certain they are not harming anyone who is not allowed to be harmed, because killing one innocent person during "jihad" could result in eternal hell. Thus jihad has big risks for anyone doing it. In fact the first person to be thrown into the hellfire is someone who fought doing jihad until they died, but they fought so that people would call him brave and didn't do it sincerely for the sake of Allah. Muhammad pbuh taught that a Jihadi will go to hell for being insincere while following the rules, so what would happen to a Jihadi who broke the rules? Breaking the rules of Allah concerning bloodshed leads to hell.

When Muhammad pbuh conquered Arabia he didn't slaughter the non-Muslim inhabitants, he let Jews and Christians practice their religion without harassment, this is a historical fact. The peace Jews were allowed to live in actually led to betrayal from the Jews of Medinah. The Jews actually tried to bring about a change in the balance of power in the city scheming with the enemies of the Muslims and shared secrets with those at war with the Muslims. In regards to people of other faiths, there was an instance pagans were on a pilgrimage without weapons and utterly defenseless, the companions of Muhammad pbuh wanted to capitalize on the opportunity and slay them easily while they were unarmed, but Muhammad pbuh would not let them and clarified that he wasn't sent with that kind of mission. When the Muslims conquered Egypt the Catholic rulers were oppressing the Copts with exorbitant frivolous taxation, after the Catholic oppressors ran away the treasury was found full of Coptic money. The Muslims returned the stolen property and let the Coptic Christians practice their religion, and they still practice their religion in Egypt to this day 1400 years later never once having been forced to convert. In fact as I write there are more Coptic Christians in Egypt than there are Muslims in Palestine and that's based on the lowest number currently given for the number of Coptic Christians in Egypt. The Egyptian Coptic church itself says they have more Christians in Egypt than France and the U.K. combined have Muslims, but Westerners claim the Coptic church inflates their numbers. Which I could understand if it was Muslim statistics but Christians don't even believe the numbers given by the Christians living in Egypt themselves. When Umar bin al-Khattab, the second leader of the Muslim State after Muhammad pbuh, accepted the surrender of Jerusalem from the Christian Patriarch "Saint" Sophronius in 637 CE the Umayya Covenant was enacted in Jerusalem. This Covenant guaranteed the Christian population the liberty to practice their religion in exchange for paying a tiny fee that would support the defenses necessary to protect the city; it wouldn't be fair for the Muslim Army to protect the Christians in case the city came under attack if they had not contributed to the defense in any way. Also Jews were once more allowed to live and worship in Jerusalem after having been exterminated in 629 CE by the Christian Byzantines. This particular stipulation of Umar allowing Jews to exist in Jerusalem was strongly protested by the Christians because they hated the Jews. They did not want any Jews anywhere near Jerusalem. This was partially because of religion and because of history. Decades earlier the Jews revolted against the Byzantine empire and helped the Persians conquer the middle east and the Jews joined the Persian army when it took Jerusalem in 614 CE. The Christians revolted once the Jews and Persians were ruling over them. When this happened the Jews and

Persians worked together to destroy the Church of the Holy Sepulchre. Then the Jews made the Christians an offer acting as a mediator between the Perians and Christians. The Jewish people said "*If ye would escape from death, become Jews and deny Christ; and then ye shall step up from your place and join us. We will ransom you with our money, and ye shall be benefited by us.*" Despite the offer the Christians did not want to become Jews, therefore the Jews and Persians killed between 57,000 and 66,500 Christians and they enslaved the 35,000 Christians who were left. So that's partially why the Christians killed all the Jews when they retook the city in 629 CE. Thus 8 years after the Christian revenge in 637 CE with the Muslims being the new rulers of Jerusalem the Christians did not want the Jews around because of their mutual hatred for each other. Yet Islam did not permit such a policy so Umar made a deal with the Christians assuring them that they could have their own separate Christian quarter in Jerusalem where no Jews would live amongst them, as per their request. Also keep in mind how when the Jews first took the city they killed the Christians since they refused to convert, when the Christians took the city back they killed the Jews since they refused to convert. The translated text of the treaty when Muslims took Jerusalem is as follows:

"*In the name of God, the All-Compassionate, the Especially Merciful. This is the assurance of safety which the slave of God, Umar, the Commander of the Faithful, has given to the people of Jerusalem. He has given them an assurance of safety for themselves for their property, their churches, their crosses, the sick and healthy of the city and for all the rituals which belong to their religion. Their churches will not be inhabited by Muslims*

44

and will not be destroyed. Neither they, nor the land on which they stand, nor their cross, nor their property will be damaged. They will not be forcibly converted. No Jew will live with them in Jerusalem. The people of Jerusalem must pay the Jizya like the people of other cities and must expel the Byzantines and the robbers. Those of the people of Jerusalem who want to leave with the Byzantines, take their property and abandon their churches and crosses will be safe until they reach their place of refuge. The villagers may remain in the city if they wish but must pay Jizya like the citizens. Those who wish may go with the Byzantines and those who wish may return to their families. Nothing is to be taken from them before their harvest is reaped. If they pay their Jizya according to their obligations, then the conditions laid out in this letter are under the covenant of God, are the responsibility of His Prophet, of the caliphs and of the faithful."

Keep in mind this treaty was just for Jerusalem, each city had custom conditions as circumstances dictated. The bit about the Christian buildings and such in the text doesn't fully explain how despite Muslims not being allowed to destroy any Christian churches the Christians were not allowed to build more churches either. What was there then would remain as long as the Christians maintained it, but no further expansions would be done and the Muslims would not destroy their stuff either, time would peacefully do that if the Christians all chose to become Muslims or didn't maintain their facilities. There is architectural evidence testifying to this tolerance which Muslims exercised. When it was time to pray the noon prayer in Jerusalem, Umar was personally invited by the Christian "Saint" Sophronius to pray in the Christian church of the holy sepulchre. To which Umar politely declined explaining that he feared if he prayed inside it then future generations of Muslims might use that as an excuse to turn the Christian church into a masjid in violation of the treaty. Thus he prayed outside the Church of the holy sepulchre and afterwards the Masjid of Umar was built on the spot, of which both places still remain to this day with the Masjid of Umar directly across the street from the famous Church of the holy sepulchre. These are verified historical facts that prove Muhammad pbuh and his companions tolerantly permitted Jews and Christians to practice their religion in peace as long the Jews and Christians didn't plot against the Muslims, betray them, preach to them, insult Islam, or make mischief in the land. The Christians even asked Umar not to let the Jews into Jerusalem, but he let them in despite the Christian requests to forbid them. Yet not only did Umar allow the Christians to keep their Church but Umar cleaned the area of Masjid Al-Aqsa which was and is the Jewish Qibla or prayer direction, as well as the Masjid which Muhammad pbuh prayed in during his night journey to Jerusalem. Originally Adam pbuh built Masjid Al-Aqsa, then it was rebuilt by Abraham and Jacob pbut, then David pbuh started to rebuild it and Solomon pbuh finished it. Then Ezra pbuh rebuilt it after the Babylonian destruction, then the Greeks turned it into a pagan temple. Hasmonean Jews recaptured the area in 164 BCE remaking it into a masjid, then Rome took control of it in 64 BCE. In 70 CE, out of hatred for Jews and their revolution the Romans destroyed the "Temple Mount" which is also called Masjid Al-Aqsa and turned it into a garbage dump so that every time any Jews in the world faced it when praying they would be facing a big pile of garbage. Over the centuries even Jews would throw their garbage on their own former "Temple Mount" because it was the garbage dump and it was the law to put their garbage there. From 614-629 CE the Jews still put their garbage there because they no longer considered it sacred, even though they faced it when they prayed. Yet since Allah declared it to be a masjid and promised it to Muslims in the Quran, when Jerusalem was conquered by the Muslims Umar cleaned the garbage dump up with his own hands and built Masjid Qibly inside the Masjid Al-Aqsa compound. Many confuse Masjid Al-Aqsa with the Dome of the Rock or with Masjid Qibley built by Umar, but the Masjid Al-Aqsa covers a very large area. The Dome of the Rock did not exist in the time of Umar bin Khattab, there was just the Rock. The Dome structure was built in 691 CE by the Ummayad Khalifh Abdul-Malik ibn Marwan and the golden tiles were only added in 1993 CE. The area of Masjid Al-Aqsa has several masjids inside it's compound and even a burial ground. The reason many think the Dome of Rock is the Masjid Al-Aqsa is because the Israeli government wants to destroy Masjid Al-Aqsa entirely and build a new Temple on the site for a secular ultimate global interfaith religion which essentially combines every religion into one. Therefore the Israeli State promotes a propaganda campaign to misinform people that only the Dome of the Rock is the Islamic holy site so they can start stealing the rest of the Muslim land and de-Islamifying it, sadly some ignorant Muslims even fall for this trap. Such claims are made so that when the Zionists damage things inside Masjid Al-Aqsa which are not the Dome of the Rock Muslims don't fight back as much. This is because currently Zionists fear the backlash if they do their demolition because all Muslims might wage Jihad in response. The famous "Wailing Wall" is simply the Western Wall of the Masjid Al-Aqsa compound and it is also Muslim property which Jews have tried to buy since the 1830s CE but Muslims said NO. After Britain conquered Jerusalem the Jews again tried to annex the wall but in 1928 CE Britain issued a white paper stating "*The wall is also part of Haram al-sharif, as such it is holy to Moslems. Moreover it is legally the absolute property of the Moslem community, and the strip of pavement facing it is waqf property*(belonging to all Muslims as a trust), *as is shown by documents preserved by the Guardian of the waqf.*" In 1967 CE the day Israel took Masjid Al-Aqsa they hoisted the Israeli flag over the Dome of the Rock, just like how the Crusaders put a cross on top of it the day they took the Masjid. Immediately the Jewish soldiers wanted to blow it up with explosives while they had the chance before any politicians could stop them, but their commander warned them not to do so because the reaction of Muslims might make them lose the whole state. They were told that Muslims were too emotionally attached to the site for them to blow it up, yet. Since then the campaign of relabeling Masjid Al-Aqsa as the Dome of the Rock has been waged so the rest of the compound loses value in the sight of Muslims. However despite not being able to blow up the Dome of the Rock, in 1967 CE the Jews did manage to bulldoze the Maghribi Quarter in front of the Buraq Wall (Wailing Wall) destroying 2 masjids and 135 homes leaving 650 Muslims homeless so the Jews could make space for tourists in front of their "Wailing wall". Many don't know that's how the empty courtyard in front of the popular "Wailing Wall" came to be, Muslims used to have homes and masjids there but the Jews illegally bulldozed them in 1967 CE. In 1984 CE the wall was registered as "property of the Jewish State", which is illegal theft but that's Israel so what can you do? Currently the state of Israel is digging tunnels under the Masjid to weaken the support structures of the compound, especially the masjids so they are vulnerable during earthquakes. This tunneling has even caused Muslim buildings to collapse, others get cracks and other structures even move as a result until they are buttressed. Other Israeli groups propose plans to blow up the Dome of the Rock and blame it on Muslim extremists or say a stray rocket hit it or just do it publicly and take full credit since some Jews have religious beliefs thinking it's destruction is necessary for their version of the Messiah to come. Chaim Weizmann of the Zionist Organization said in 1918 CE that Jews go to the Wailing Wall to *"bewail the destruction of the Temple and pray for its restoration."* Which means when Jews go to that wall they are praying for the Masjid Qibly and the Dome of the Rock to be destroyed and have their own Jewish or Interfaith Temple built on the ruins. Some of these Zionist Jews even go to America and tell Christians that the only thing stopping Jesus pbuh from coming back is the existence of the Dome of the Rock. Some Zionists claim a "Third Temple" is needed, with some proposing a Jewish Temple or having it be run by the yet to be developed Interfaith global composite religion. Such inflammatory things are preached publicly by Jews and Zionists and it's allowed because of America's support for Israel. Yet religious persecution doesn't stop with comments or covert plans/actions, it turns into public violence. Many incidents have taken place where Jews or Zionists just come into Masjid Al-Aqsa and start killing Muslims, while the Israeli soldiers watch or join in saying

49

those Jews are killing in "self-defense". This harassment of Muslims inside their own holy places isn't reported on the mainstream news because the news outlets believe the Israeli State when it says the Masjiid Al-Aqsa is "only the Dome of the Rock" and America supports the state of Israel. Plus it's Jews, what news outlet would dare publicly declare Jews to be religiously persecuting people in their sacred places of worship? Didn't you hear of Hitler and the Holocaust? If people knew that's what goes on in Israel that might make America look bad for being such a close and supportive ally. Muslims even get harassed and physically assaulted at the Dome of the Rock by Israeli soldiers and Jewish settlers and such abuse happens in front of tourists too. Yet most tourists are religious non-Muslims who support Israel so they don't really care to do anything about it. On April 11, 1982 CE a Jewish American named Alan Henry Goodman went into the Dome of the Rock and shot those inside killing 2, wounding 30. He spent 15 years in prison before receiving a full pardon by the Israeli state prior to being released. Muslims are also prevented from entering the Masjid Al-Aqsa compound and the Israeli State frequently bans all male Muslims who are under 50 from entering the area. Also the state of Israel forbids some masjids from calling the athan, frequently without even giving them a reason for the bans or saying that the athan is "noise pollution". The Jews don't even let the Muslims call each other to pray in the holy land because they don't want the call to be heard, nor the prayers to be made. Israeli heads of state and pro-Israel supporters even publicly claim that *"There is no such thing as the "Palestinian people". The mythical Palestinian people never existed."* Even though they've existed for thousands of years and are an endangered species actively being exterminated at this very moment. Regardless of this shameful sinful treatment in modern times, yesterday, today and until the last day all Muslims will still abide by the treaty of Umar regarding Jerusalem which he made in 637 CE. Although currently the Zionists control Jerusalem and oppression exists, for now. Yet that's why Allah taught Muslims how to wage Jihad. However Umar's treaty is still the Muslim standard for Jerusalem, when Muslims say we want to rule Jerusalem again that is what our policy will be. In modern conditions one would just have to consider the word "Byzantines" to be "Zionists" and since the Jews today would likely fear reprisals for their harsh treatment of the Christians similar to how the Christians hated/feared the Jews when Muslims first took Jerusalem, one might also have to consider the word "Christians" in Umar's treaty as though it said "Jews" in a modern rendition and vice versa. The Muslims first conquered Jerusalem in 637 CE when Christians surrendered after a 4 month siege, the armies never even fought each other. There were some deaths because they shot arrows at each other but no actual hand to hand combat took place. The reason they never had any melee battle was due to religious prophecy. The siege of Jerusalem took place about a month after the Muslims won the Battle of Yarmouk in Syria. For the first 3 days nothing happened, the Muslims surrounded the city but no warfare and no dialogue between the two sides took place, it was very odd. On the 4th day the Muslims sent a messenger to inform the people of Jerusalem they had 3 options: Embrace Islam, Surrender and agree to pay Jizyah or warfare. The Christians said they would never

51

embrace Islam and preferred death. Then the Muslim army besieging Jerusalem wrote to Abu Ubeidah their commander for permission to attack since he had only given them permission to besiege. After given orders to attack then arrowfire was exchanged between Muslims on the ground and Christians upon the ramparts for 10 days with a few getting killed in the process. On the 11th day after fighting began the commander of the Muslim army Abu Ubeidah arrived so the Christians then asked Sophronious what to do. He swore by "the Injeel" that if their commander has come then their end is at hand because the knowledge passed down since the time of Jesus pbuh said that a reddish brown man named Umar who is a companion of Muhammad pbuh would conquer Jerusalem. So the Patriarch of Jerusalem got the priest together with bibles and crosses and went to inspect the Muslim general to see if he was the one prophesied to conquer Jerusalem. They told the Muslims in Arabic: *"We know the description of the man who will conquer our city and all the land. If he is your commander then we shall surrender otherwise if it is not him we shall never surrender."* So Abu Ubeidah approached the walls, when they saw him the Christian priests and scholars said that's not him and exhorted the population to keep fighting. Thus the arrowfire continued to be exchanged everyday over the course of 4 months. Then the Christians again begged Sophronious to do something and see what the terms of the Muslims were and if they were agreeable then they'd surrender because they were weary of fighting and help was not going to be sent to them. So through translators Sophronious and Abu Ubeidah spoke as follows:

Sophronious: "*What do you want in our sacred city? God becomes angry and destroys whoever targets it.*"

Abu Ubeidah: "*Yes, she is a indeed a noble city. From her our prophet was taken to the Heavens and approached his Lord the distance of 2 bows or even closer. She is the origin of the Prophets and their graves are in her. We are more entitled to her than you and will stay here until Allah grants her to us just as He granted the other cities to us.*"

Sophronious: "*So what exactly do you want from us?*"

Abu Ubeidah: "*One of 3 options- the first is that you say "There is no deity but Allah alone who has no partner and that Muhammad is his slave and Messenger." If you accept these words then our rights and duties will be the same.*" (Note equality comes with option number 1 only. Equality is not part of option 2 or 3.)

Sophronious:"*Those are very great words. We already recite them except that we do not accept Prophet Muhammad to be a Messenger.*"

(Note that Sophronious referred to Muhammad pbuh as "Prophet Muhammad" yet he didn't accept him as a messenger of Allah, he didn't say it wasn't true but just that they didn't accept "Prophet Muhammad" to be a Messenger. Those are the Christian "Saint" Sophronious' own words, not mine. He as a learned Christian scholar and leader called Muhammad pbuh a prophet when interacting with Muslims but just said he and the Christians who knew he was a prophet didn't accept him.)

Abu Ubaidah: "*O enemy of Allah, you lie! You people have never taken Allah to be One. Allah has informed us in his Book that you say "Christ is the son of God" but there is no deity except Allah. He is Pure and High, incomparably high above what the oppressors attribute to him.*"

Sophronious: "*Then this option we cannot accept. What is the second option?*"

Abu Ubaidah: "*You will surrender your city and pay us the Jizyah in a state of humility just as the other cities in Syria are paying.*"

Sophronious: "*That is even worse than the first option because we do not lower ourselves to anyone.*"

Abu Ubaidah: "*Then we will fight you until Allah grants us victory over you. We will enslave your women and children and kill all the men*(of fighting ability) *who oppose the declaration of Allah's Unity and cling to their declaration of disbelief.*" (This is called option 3, non-Muslims should not take option 3. I'd suggest option 1, but option 3 is never pretty for non-Muslims because people get killed and enslaved. But keep in mind this is a military commander giving these options. A regular Muslim today can't offer such options because they do not have the authority and the circumstances for such negotiations and proposals rarely arise in the life of most Muslims. Wars are declared and treaties are enacted by the rulers of Muslims, not unilaterally by individuals. (like Osama bin Laden who had no authority to declare war on the West) Thus for Muslims reading this, know that you will likely never ever be able to give people these options, because it is a type of diplomacy to be done between diplomats not the masses. So don't go to some random disbeliever and say they have 3 choices because you cannot just go engaging in national diplomacy when the Muslim Ummah hasn't given you that job. That was Abu Ubaidah's job, a regular Muslim's job today is to politely offer disbelievers option 1 and give them many reasons to pick it before they die of non-related causes or ideally before a qualified diplomat gives them the 3 choices(which hasn't happened in a very long time).)

Sophronious: *"We shall not surrender our city even if it means death. Why should we surrender when we have ample provisions to withstand your siege, excellent equipment and powerful soldiers? We are not like those cities which voluntarily paid Jizyah. They are people upon whom Christ is angry, hence he placed them under your rule. However, we live in a city in which if anyone prays to Christ, He answers him."*

Abu Ubaidah: *"You lie again, O enemy of Allah."* then Abu Ubaidah recited the verse of the Quran 5:75 which means: ***"Christ, son of Mary, was nothing but a Messenger. Many Messengers passed before him and his mother was a truthful, virtuous lady. Both of them ate food(which Allah does not do)."***

Sophronious: *"I swear by Christ! Even if you besiege us for twenty years you will never be able to conquer us. We will only be conquered by a certain man who is described in our Scriptures and knowledge. This city will be conquered by a companion of Muhammad called Umar. He will be known as al-Faruq (the one who distinguishes between Truth and Falsehood.) He is a stern faced man who is not concerned about the rebuke of people when it comes to obeying Allah. This is not your description."*

55

Abu Ubaidah: (laughing) *"By the Lord of the Kaba! We have conquered the city. Will you recognize this man if you see him?*

Sophronious: *"Why should I not when I have his exact description and age?"*

Abu Ubaidah: *"By Allah! He is our Khalifah and Sahabi of our Prophet pbuh."*

Sophronious: *"If he is as you say, then seeing that you know us to be truthful, you should stop the bloodshed and send for your man to come here. If we see that it is in fact him, we will open our city for him without any disturbance and pay the Jizyah. "*

Thus the Christians and Muslims agreed to a ceasefire until Umar bin Khattab came in person to be inspected by the Christians. However at the time Umar was in Madinah, Arabia and transportation was not very fast. Thus the Muslims decided to have another Muslim named Khaalid bin Waleed pretend to be Umar because he looked the most like Umar out of everyone present. So in the morning the Christians were told that "The Khalifah has come."

Abu Ubaidah then came with Khaalid and said to the Christians: *"The man whom you seek has come."*

Sophronious said: *"Tell him to come forward so that we can see him."*

After Khaalid came forward Sophronious said: *"By Christ! This looks like it is him but some signs are missing. I implore you in the name of your religion, tell me the truth-who are you really?"*

Khaalid said: *"I am one of his companions."*

Sophronious said: "*You Arab boys, this is treachery! By Christ! Until we do not see the described man we will neither open the gates nor will any of us speak to you even if you besiege us for twenty years.*"

Then the Christians shut the gates and refused to talk to the Muslims anymore after that, so the Muslims wrote Umar telling him what happened asking him what they should do. Umar asked for advice and decided he would come to Jerusalem himself. After many days Umar had finally arrived in person.

Abu Ubaidah called out to the Christians: "*O people of this city, the Commander of the Believers has arrived. What will you do about what you had previously said?*"

Sophronious came in special Christian attire with clergymen carrying a special cross displayed only for festivals(it was a Sunday). The governor Batlic was with him and advised "*O Father, you should be able to recognise him with certainty. Failing that we will not open the gates and will fight until either we or they are destroyed.*" Sophronious told Batlic "*I will do that.*", then he shouted to Abu Ubaidah: "*What do you want, old man?*"

Abu Ubaidah replied: "*This is Umar, Commander of the Believers. There is no commander above him. Now come to receive his amnesty, to surrender and to pay the Jizyah.*"

Sophronious calmly said: "*O man, if he is really the highest ranking then let him approach us so that we can recognise his characteristics. Let him come out alone from amongst you and stand directly in line with us so that we can see him. If he is the man described in the Scriptures then we will come down and seek*

57

amnesty and pay the Jizyah. If it is not him then you will get nothing but battle from us."

After Umar bin Khattab presented himself Sophronious rubbed his eyes, looked repeatedly and loudly shouted out: *"This is he who is described in our books, the man who will conquer our city without doubt."* Sophronious quickly ordered the Christians of Jerusalem: *"Woe unto you! Go down for amnesty and protection. By God! This is the mentioned companion of Muhammad bin Abdullah."* (Of which the eerie part is that the Muslim army had never told Sophronious that the name of Muhammad's father was Abdullah. So how did the Christian "Saint" and scholar of Jerusalem, Sophronious, know the correct name of the father of Muhammad pbuh? It must have been in his "Scriptures" because it was a rare and obscure piece of information not widely known to people who didn't grow up or meet Muhammad pbuh personally. This was because Muhammad's father had died before he was even born. So Abdullah the father of Muhammad pbuh had died over 70 years earlier, yet somehow Sophronious knew his name and that he was Muhammad's dad. Furthermore from his order one can see that the Christians of Jerusalem also must've known something about someone called Muhammad bin Abdullah, because Sophronious wouldn't have told the Christians that full name unless there was a reason to say that exact name to them because it meant something very important and would get them to obey.) According to the Christians of Jerusalem, Umar bin Khattab perfectly matched the description of the man who Christians said was destined and prophesized to conquer Jerusalem in their

Scriptures and ancient knowledge, so they surrendered as I had previously explained. However some Christians/Romans did not like the idea of being ruled by Muslims and they made a plan to attack the Muslims while they were weaponless saying their daily prayers. One of them named Abu al-Jaid who heard of the plan said: "*O people do not do this. Do not betray them. If you are going to do this I will inform them.*" So they asked him what they should do. Abu al-Jaid told them, "*Display all your worldly goods to them. Whoever sees them will not be able to control himself. If they then seek to seize it wrongfully then you will have an excuse to do as you please.*" So the Romans did just that purposely showing off all their material wealth hoping a Muslim would try to take some so then they could attack the Muslims and kill them all claiming it was self-defense against oppression. The Muslims had never seen such wealth before but none came near it or even touched a single piece of it. The Muslims just said: "*Praise be to Allah who has granted us control over the houses of people such as these. If He had to regard the world equal to the wing of a mosquito He would not have granted a disbeliever a drop of water to drink.*" After seeing the restraint of the Muslims in the face of such temptation Abu al-Jaid remarked: "*These are the people whom God has described in the Torah and the Injeel. They will remain on the Truth and none will be able to approach them for as long as they remain like this.*" Historically it was one of the most peaceful and amicable conquests of Jerusalem ever and it was all because of religious reasons. Devout strict Christians actually surrendered Jerusalem to the Muslims in 637 CE because their religious "Scriptures" said so. Now obviously Umar and the name of Muhammad bin Abdullah is not mentioned in the bibles sold today. So it makes one wonder what "Scriptures" these devout Christians in Jerusalem during the 600s CE were reading and talking about? They themselves said it predicted the "Prophet Muhammad ", whom while they didn't accept him as a Messenger they believed he was a prophet and that his companion Umar was divinely decreed to conquer their sacred city according to their religious "Scriptures" and ancient knowledge from the days of Jesus pbuh. That is the actual history of how and why Muslims conquered Jerusalem in 637 CE. Now why is it that nobody in the West ever gets taught how and why Muslims were able to have Jerusalem surrender? Honestly if you ask a non-Muslim if they know when Muslims first conquered Jerusalem and how, they draw a big fat blank with no information about it. They will know about the Crusades and when that was and all kinds of stuff about that, but if you ask for info about when the Muslims first took the holy land from the Christians they generally don't know anything at all and will just assume it was a bloody conquest by the sword via Jihad and the Crusaders had to defend the persecuted Christians. I had even considered myself an expert on the Crusades and I had no idea how, when, why or who the Muslims were when they first conquered Jerusalem. To my crusader mentality all that mattered was that Christians didn't have it so they should just kill whoever had it that wasn't Christian. If anybody non-Christian had it then I thought they should be fought and killed and didn't want to know anything about their religion because I thought it must obviously be evil and false because it wasn't Christianity. Anyone non-Christian was on my kill list and I didn't feel I needed to know their faith or history because I was guided by the bible, Jesus, God and the Holy Spirit and I could feel I was right and I felt so darn right too. Yet the question still remains why don't Christians get taught about how and when the Muslims first conquered Jerusalem in schools and the media? It truly is a remarkable religious and historically important event that is extremely relevant to modern day turmoil and politics. So why don't Western people get informed of this? It is because everyone would wonder what happened to these Christian Scriptures and prophecies that the Christians who lived in the sacred city of Jerusalem firmly believed in. Obviously later Christians would not want such Scriptures to become mainstream, especially after the stuff prophesied in them was perfectly fulfilled by Muslims. So to learn about the conquest of Jerusalem would be to learn that the Christians in Jerusalem had Scriptures they thought had divine instructions specifically regarding Muhammad and Muslims, but they don't exist today. Why don't they exist today? Why aren't these ancient Scriptures the Christians of Jerusalem believed in present in the bibles which Christians today believe in and zealously call Scriptures? I mean just imagine if Muhammad pbuh was mentioned by name in the bible Christians are reading today in a positive light, what do you think would happen? Christians would be inclined to research Islam and possibly become Muslims just as I did. So did Muslims just make up this story about the Conquest of Jerusalem? No, it is the authentic historical account and is documented by all historians, the letters sent to Umar exist and the results are factual. There is no other alternative version to it. Those Christians from Jerusalem surrendered their city to Muslims based on those Scriptures and what they said, but for some reason those ancient religious Scriptures aren't read by Christians today and Christians say the Scriptures don't say what the Christians in the 600s CE said their Scriptures said. So who are we to believe? The Christians of the 600s CE saying and writing and surrendering entire cities based on their claim and belief that their Scriptures talk about Muhammad pbuh, or the Christians of the modern era thousands of years and edits later saying their freshly printed edition of the bible doesn't? Which group is more likely to be lying or mistaken about what Jesus pbuh taught and have corrupted/missing religious texts? Since Christians cannot handle the truth and what it implies about Islam and their "Scriptures" they ignore it and refuse to teach the world about when, how and why Muslims first captured Jerusalem. If anything they give dates with no details so a bigot draws their own fantastical fanatical false assumptions. Regardless of what one thinks of the conquest of Jerusalem by the Muslims in 637 CE it was religious in nature and design. It was peaceful without persecution or genocide. Such is an Islamic conquest, that is if/when Jizyah is accepted, if Islam is accepted then there is no conquest but equality and teamwork as one Muslim nation until the Day of Judgment.

462 years after the first Muslim conquest, when the Christian crusaders conquered Jerusalem they slaughtered nearly everyone inside Muslim, Jew, man, woman, child (and I only say nearly because a few escaped the slaughter and fled to tell Muslims in other cities so they could prepare themselves). Yet if you read the Christian records, according to the Crusaders there were no Muslim survivors. So

Muslim records actually make the Crusaders seem better than the Christian records do. As unlikely as it seems today, Jews in Jerusalem actually fought alongside the Muslims against the Christians because the crusaders came with the slogan of *"Embrace the Cross or die!"* Pope Urban II misinformed them that all the Christians had been slaughtered by the Muslims and prohibited from practicing their religion when motivating them to crusade, despite the Christian Hospitaler knights having operated a hospital in Jerusalem since 1050 CE. Thus the crusaders decided that they would kill as many non-Christians as they could. It is reported that when the crusaders captured Jerusalem the streets were flooded with blood that reached up to the knees of their horses. The Jews were burned alive in their synagogue. After the crusaders entered Jerusalem, on the 15th of July 1099 CE more than 70,000 dead bodies of Muslim children and women were found just in the Qibley Masjid alone. The Dome of the Rock had a cross put on it and was turned into a church, the Qibley Masjid became a palace and the rest of Masjid Al-Aqsa became stables for the crusader horses, later a tiny portion of space was allotted to the Knights Templar to be their headquarters next to the Dome of the Rock. The few Muslims who survived the initial bloodbath when discovered were brought to the Masjid Al-Aqsa and would be killed via a long agonizing crucifixion on a large cross. Muslims were then banned from being in all cities under crusader control and got turned into slaves to work the land now owned by the crusader knights. Most masjids got demolished except for the ones which were converted into Christian churches or palaces. The athan would not be called in Jerusalem for the next 90 years. But why did this brutal invasion happen if Christians had peacefully lived in Jerusalem under Muslim rule and been allowed to practice for over 400 years? Keep in mind I'm referring to the Muslim rule, not Shia rule. You see at the time the middle east was even more divided than it is today with nearly every city being almost like it's own nation state. Honestly if you think Muslim nations today are weak and divided, back then they were even weaker and more divided than they are now. Weak as they are Muslim nations today are actually stronger than they were during the first crusade. The Khilafah was a shell of it's former self because of Muslim disunity and infighting, which allowed Shia to gain power in North Africa and spread as the Fatimid empire reaching from Algeria to Palestine and Yemen. In 1009 CE the Shia ruler Al-Hakim, who was leader of the Fatimid empire, destroyed the church of the holy sepulchre in Jerusalem and persecuted Christians. Then in 1016 CE this Shia, Al-Hakim, claimed he was the incarnation of God on earth and founded a cult within Shiism that became what is known as Ismai'ilism. One night while wandering alone in the hills, Al-Hakim went missing and was never heard from again; leading Ismai'ilis to further split off into different sects in doctrinal deviance like the famed Assassins of Masyaf who taught a doctrine later adopted by the Bavarian Illuminati. After this tragedy of Shia oppression, the church of the holy sepulchre was nearly completely rebuilt by 1048 CE except for it's western parts which were still ruins when Christian pilgrims returned to visit the city. Eventually the Seljuk Turks recaptured Jerusalem in 1073 CE from the Shia Fatimids. But rebels took over Jerusalem in 1077 CE which led to the Seljuks reconquering the city, with more drama in 1091 CE when the governor of Jerusalem died and his 2 sons disputed with each other. Meanwhile the Muslim Seljuks were conquering Byzantine Turkey ever since the famous battle of Manzikert in 1071 CE, where the acting senior Byzantine Emperor Romanos IV, who had married the widowed empress, was captured by the Muslim army. When they met, the commander and Sultan Muhammad bin Dawud Chaghri, also known as "Alp Arslan", asked the Emperor *"What would you do if I were brought before you as a prisoner?"* Romanos IV said, *"Perhaps I'd kill you, or exhibit you in the streets of Constantinople."* to which Muhammad bin Dawud Chaghri said, *"My punishment is far heavier. I forgive you, and set you free."* After this the same exact terms of peace which the Muslims offered the Byzantines before the battle took place, was again offered to the Byzantine Emperor. Muslims always have the same terms, either accept Islam, pay Jizya or war until you pay Jizya or accept Islam. Or on rare special occasions we will have a truce for a maximum of 10 years, according to the example of Muhammad pbuh. After losing the battle Romanos IV remained a prisoner for a week, during which he ate at Muhammad bin Dawud Chaghri's table whilst concessions were agreed upon. The territories of Antioch, Edessa, Hierapolis, and Manzikert were surrendered. When a payment of 10 million gold pieces demanded by the Muslim commander as a ransom for the Emperor was deemed too high by the Emperor, Muhammad bin Dawud Chaghri reduced it asking for 1.5 million gold pieces as an initial payment, followed by an annual sum of 360,000 gold pieces. Next Muhammad bin Dawud Chaghri married his son to the Byzantine Emperor's daughter and gave the Emperor many gifts and sent 2 escorts to help protect him on his way back to Constantinople. That's how the infamous Muslims treated the Byzantine Emperor when he was a prisoner of war while Muslims were conquering Byzantine territory. However months after Romanos IV returned he gave co-emperorship to Michael VII the rightful heir, since by that age he was old enough and suddenly interested in being Emperor. Michael VII's sudden interest was likely due to court politics and pressure since prior to the battle of Manzikert he had no interest in politics or governing. Soon after this Romanus IV was overthrown and agreed to live out his days in a monastery. Yet prior to his final defeat and imprisonment by his own Byzantine countrymen and co-religionists he sent Muhammad bin Dawud Chaghiri all the money he could collect with a letter saying *"As emperor, I promised you a ransom of a million and a half. Dethroned, and about to become dependent upon others, I send you all I possess as proof of my gratitude"*. The Byzantines then blinded Romanos IV and prevented him from getting medical care in the monastery, which eventually led him to die a slow painful death as a result of infection. Michael VII decided not to honor any agreements which had been made with the Muslims and his successors hated Muslims even more than he did. In 1092 CE the Seljuk Empire fractured after the death of the Sultan when the four sons of the Sultan and his brother all fought each other carving out their own mini states. In the midst of this mess in 1095 CE Pope Urban II gives his famous speech telling the Christian world to kill Muslims launching the first crusade claiming they were persecuting Christians preventing them from visiting Jerusalem and destroyed the church of the holy sepulchre.

Whereas in reality it was the Shia who did that 86 years earlier. Then ironically in 1098 CE, 3 years after the first crusade began and a year before the crusaders took Jerusalem, the Shia Fatimids recaptured Jerusalem from the Abbasid Khilafah, who had it gifted to them by the Seljuks, and the Shia expelled the Christians again. When the crusaders arrived at the gates they met the Shia and were completely clueless that their crusade against the Muslims throughout Seljuk Turkey directly helped the Shia oppressors from Egypt retake Jerusalem and oppress Christians, which the Shia expelled just before the crusaders got to Jerusalem thereby acting as a confirmation for the lies the crusaders had been told about Muslims. So the whole problem comes down to the West believing Shias are Muslims when they aren't, then as a result attacking Muslims because of their hatred for Shias and the lies their Christian leaders told them. In short, Shias started oppressing Christians and the Muslims did jihad and stopped them, then when the problem was nearly resolved and forgotten with the wounds healed, the Pope comes and blames Muslims for having caused a problem they were innocent of and the Pope orders that the Muslims are to be killed. So crusaders attack the Muslims because of what they were told about Muslims even though the Shia were the real bad guys, and by attacking the Muslims they helped the Shia criminals regain power so they could oppress again so that when the crusaders see the Shia oppression upon reaching Jerusalem they thought the Pope was right. Upon defeating the Shia, they say, "*See, we told you those Muslims were evil and God is on our side*". Meanwhile most Western crusaders never looked into Islam or what actually happened and just believed they were the good guys because the Pope told them so and they "had a feeling" along with circumstantial "evidence". Thus they fought Muslims because they mistake Shia for Muslims and equate them. So then when they later unjustly oppressed and murdered Muslims and the Muslims did jihad and fought back, the crusaders didn't realize that they were actually evil oppressors who were, are and to this day continue to be in the wrong. But was it just a simple mistake by Pope Urban II or did he know the Muslims were innocent and that the Shia were of a different religion? Unfortunately it doesn't appear to be accidental. Politically the first crusade was designed to aid the Byzantine emperor Alexios I, who's army was too weak to fulfill his desire to recapture the land his ancestors had once brutally ruled over, which the Muslim Seljuks had recently liberated. Past Shia actions simply provided Pope Urban II the excuse the Church was looking for to attack Muslims, that's why it didn't matter that the crimes used to justify the crusades were done 86 years beforehand or by people who weren't even Muslims. To show how wrong the first official crusade was imagine if the leader of the U.N. today said "*86 years ago X building was destroyed in Y country and innocents were killed by S people. Therefore even though X building has been rebuilt and Y country is now ruled by M people let's invade Y country and kill all M people!*" This was what Pope Urban II did and it worked because he didn't say anything about 86 years ago or the difference between S people(Shia) and M people (Muslims). He knew better though and was 86 years behind when it came to world politics. It was because there was no oppression or injustice under Islamic Shariah in any of the

Muslim territories that the Shia crimes of the distant past had to be cited as reasons to motivate crusaders to go slaughter. This doesn't seem familiar does it? Historically that's what happened. If such things are still happening then perhaps we're still in the "Dark Ages", with our time just being a different shade of dark.

The 3rd crusade took place after the Muslim leader Salah Ad-Deen recovered Jerusalem, which the Christian Crusaders surrendered. Every historian attests that Salah Ad-Deen was much more tolerant than the crusaders. Some famous quotes of Yusuf ibn Ayubi / Salah Ad-Deen include:

- *I warn you against shedding blood, indulging in it and making a habit of it, for blood never sleeps.*

- *"If people who carry arms are faithful and sincere to their religion then intrusions from outside and the conspiracies within the state can in no way harm you. Take your eyes further than the borders. There are no borders for an Islamic Empire. The day you and this glorious religion of Islam is restricted to borders, that day you should know that you will be imprisoned in your own dungeon. Then your boundaries shall shrink. Take your eyes further than the Roman Sea. Temples cannot cease your way. Do not be afraid of the flares within your house. They will be extinguished in just one blow. Stay away from only two curses; dashed hopes and mental extravagance. A human first gets dashed hopes, then through mental extravagance takes the easy way out."*

- _Remember that no non-Muslim can ever be the friend of any Muslim._ *The Crusaders are incapable to confront us in battlefields, their tactics have failed, and that is why* _they are trying to eliminate loyalty and religion from the minds of the emerging Muslim generation._ *The weapon they are using is jeopardous. It is mental extravagance, sluggishness and pretermission. To generate all three of these dreadful elements in you, Christians and Jews have united. Jews by the means of their women are advocating animal spirit and making you drug addicts. I will not say that animal spirit and addiction will wreck your afterlife and you will make way in hell. I want to say that these dreadful elements will shape your life in this world like hell. What you think is the flavour of paradise; it is the torment of hell. You will be the slaves of Crusaders who will roam about humiliating your sisters, the pages of Quran will flutter in your alleys and your Masjids will become stables.*

- *"An oppressive leader in actuality is a weak leader. He attempts to defeat his rivals with tyranny and violence or by creating greed for wealth. "*

- *If you want to destroy any nation without war, make adultery or nudity common in the young generation.*

- *Victory is changing the hearts of your opponents by gentleness and kindness.*

- *"I fear a day will come when Arabia will be a Muslim majority area but the cross of the Christians will rule the heart of the Muslims."*

Salah Ad-Deen didn't kill the inhabitants of Jerusalem, instead he set a low price for the Crusader generation of Christians to buy their way out of slavery and freed the elderly, women, widows, children, and all who were too poor to pay and got abandoned by the wealthy Christians who simply paid their own price and left. Only one cross was broken by Salah Ad-Deen when he retook Jerusalem, it was the killing cross the crusaders used to crucify Muslims in Masjid Al-Aqsa. Most of the other Crusader cities conquered also surrendered because of the policy Salah Ad-Deen had of letting the Crusader generation of Christian civilians live and leave freely if their city surrendered. If they resisted they'd become prisoners of war to be ransomed, or enslaved, because the Crusaders rarely paid the ransoms. Whereas those Christians who could prove their families lived in the Holy Land prior to the Crusades and didn't side with the Crusaders were allowed to live in their cities as they had before the crusaders came. Also of note is that all of the known Muslim Scholars of that time participated in the Conquest of Jerusalem as soldiers in Salah Ad-Deen's army. This is because genuine Muslim Scholars practice what they preach, they don't just preach Jihad and let others wage it, if it's a legitimate Jihad they have a tendency to join in themselves. This also teaches us that when no known Muslim Scholars support a group claiming to be waging Jihad, it definitely makes one doubt whether that group is rightly guided. During the Crusading era there were approximately 80,000 Christians living in what was Muslim territory. A famous Egyptian Christian historian of the time, Abu Salih, related that he counted 707 churches and 81 monasteries in Muslim Egypt. Do you know how many Muslims and Muslim institutions were in Christian Europe at the same time? Very few. As is evident in that Adam Neuser, a German priest and the leading preacher in the city of Heidleberg who became a Muslim in the late 1500s CE, was imprisoned and had to escape twice in order to get to Muslim Istanbul before he could practice Islam, because Islam was not welcome in Europe. In comparison during 1183 CE, whilst Salah Ad-Deen was actively waging jihad against the crusaders this proclamation he made in Aleppo illustrates how the disbelievers were treated under Shariah:

"After we ordered that the dhimmis should wear the distinctive signs that differentiate them in their appearance from Muslims, and that this should be established in accordance with the requirements of manifest and pure divine law, the news reached us that a band of irresponsible hoodlums had attacked the dhimmis with despicable words and acts, curbing what the pact of protection had granted them regarding their means of subsistence and their situation. We disapproved of that, and we are obliged to prohibit such things from being said and done. We command what is set down in that document and judgment, namely: guard and protect those dhimmis; abstain from harming and harassing them; do not cause them any wrong; do not institute wrongful proceedings against them and do not deviate from the straight and narrow regarding them; do not alter the justice that is guaranteed them; do not interfere with the benefits supplied them; do not attack them in acts or words. Let them therefore not be subjected to hearing disagreeable, prejudicial, or unjust words. May they obtain the rights of the contract of protection consistent with its statutes and with justice; and may they enjoy these rights to the full in what they are granted in

benevolence by them. May their blood and inviolable property be protected; may the means of subsistence distributed to them be abundant; may the activities guaranteed them be strengthened; may the interests they wish to defend be organized and may the decrees of religious law as it is defined apply to them... The emirs and other governors must execute that judgment, must guarantee these dhimmis against injustice, and must protect them in every situation and, in case of incident from any harm and loss."

Despite this treatment Christians received under Shariah law, which is better than they are treated in Christian countries today, just weeks after Salah Ad-Deen recovered Jerusalem, before the news had even reached Europe Pope Gregory VIII launched the 3rd crusade. The crusaders had 3 main leaders: King Richard of England, Emperor Frederick Barbarossa of Germany and King Phillip II of France. The rulers were leading the armies themselves personally putting their own lives on the line for this campaign. Just imagine if the leader of your country were going to war thousands of miles away to do hand to hand combat with the enemy. This was a major military campaign of historical proportions, the likes of which has never been imitated again in military history. To raise funds for the war England and France instituted a special emergency tax of 10% on all revenue and movable property called the "Saladin tithe", while the few Jews who were too important to rob and kill had to pay 25%. This was a temporary emergency Crusade tax, meaning during the crusades most citizens were able to keep 90% of what they earned and there were no other taxes. Today most governments levy heavier daily taxes on their people than zealous Crusaders did during the "Dark Ages", when they physically forced people to pay "temporary emergency taxes". Mathematically it's cheaper to be oppressed under medieval fiefdom than to have modern "freedom". Ironically historians say that the entire Western tax system can trace it's origin to the taxes levied during the Crusades. So it makes one wonder that if taxes were introduced to post-Roman Europe and the West as an emergency measure in order to pay for the Crusades against Islam/Muslims then why do they still levy taxes today? Does that mean the Crusades haven't stopped yet? So many people "*took up the cross*" that there weren't enough boats in all of Europe to provide transportation, forcing Fredrick and his army to march on land. Frederick and his army entered the famous city of Tarsus and then history has conflicting accounts of what exactly happened next. Some say he fell off his horse and drowned, while others claim he got hit by a falling tree branch and drowned, some say he took a bath in the river and got a chill then died as a result, while still others say he had a heart attack and drowned. However all agree that Frederick died in Tarsus with the death having something to do with a river, thought to be the Saleph River. When I would read about the death of Frederick as a Catholic it seemed to be a tragedy that I couldn't explain. As a Muslim it is plain to see that God was not in support of Fredrick's cause and supernaturally killed him sending the angel of death to take his soul similar to how God killed the Pharaoh. Soon after his death, Frederick's remaining army fell sick and ill with many dying or deserting. Although Frederick had made an oath to the Christian concept of God that he would set foot in Jerusalem; thus his son wanted to have his father's oath fulfilled. The dead body of Frederick Barbarossa was boiled, stuffed and put in a barrel of vinegar

pickling him so he didn't decompose. Yet for some
inexplicable reason the body still rotted and the barrel
actually exploded before they reached the Holy Land.
About 95% of the German "holy army" never made it to the
holy land. These are historical facts. Now if we look at this
from a religious aspect and say God was either on the side of
the Crusaders or the Muslims, then it would appear based
on the humiliating demise of the Christian king and his
army that God was against him. Especially when one
considers Frederick took an oath to his version of God that
was made impossible for him to fulfill, even as a corpse. The
rest of the crusade went the same way with both Phillip II
leaving and Richard "the Lionheart" leaving to return to
England in order to save his kingdom. Yet on the return
journey Richard became bankrupt and was imprisoned
before reaching England, having to be ransomed for an
astronomical amount that bankrupted England. Crusader
historians depict Richard as a religious man, yet he refused
to enter Jerusalem when offered the opportunity by Salah
Ad-Deen. Richard said he only wanted to enter Jerusalem if
he could conquer and keep the city. This reveals Richard to
be more materialistic than religious, especially considering
how he left a "religious war" in order to keep his earthly
kingdom. Coincidentally the Muslim leader Salah Ad-Deen
died months after Richard left. Many historians think that
had Richard stayed a few more months until the death of
Salah Ad-Deen and kept fighting the world today would
have been a very different place. Although this is pointless
speculation by historians, we know that God intended
everything to turn out the way it turned out. The other
crusades also failed in unusual ways. The total number of

medieval crusades numbers in the double digits. However not all of them were directed at Muslims exclusively, some were against Jews and what were considered heretics living in Europe. The 4th crusade actually ended up with the Roman Catholic crusaders slaughtering and looting the Eastern Orthodox city of Constantinople, despite the original intentions being to conquer Jerusalem. The 4th crusade involved Christians killing Christians in the name of Christianity in Christian land. Pope Innocent III planned the 5th crusade in 1208 CE but waited until the timing was opportune in 1213 CE to launch it with a new goal to conquer Egypt before Jerusalem. It took years to gather momentum and quickly failed nearly upon arrival. The end result of the 5th crusade being the Crusaders returned the city of Damietta in exchange for all the crusaders Muslims captured in battle afterwhich an 8 year peace treaty, one of the many offered by the Muslim leader Al-Kamil, was agreed to in 1221 CE despite angry protests by the papal representative. Guess what happened next? 7 years later, with 1 year left on the peace treaty, the 6th crusade was launched in 1228 CE by the recently elected Pope Gregory IX, who was also the inventor of the Papal Inquisition. King Frederick II was the prime participant in this crusade mainly because the Pope excommunicated him for not going on the 5th crusade and said the only way to become a Christian again was to start/lead the 6th crusade. The Pope even launched a crusade against Frederick II because of his lack of will to crusade. It was a "*You're either with us or against us*" mindset. In 1229 CE Al-Kamil offered peace to the crusaders even giving them Jerusalem, Bethlehem and Nazareth, on the condition that they wouldn't rebuild any of Jerusalem's fortifications which had been demolished and that after the 10 year truce expired the territory would be returned to the Muslims. Well not only did the Christians kick the Jews out of Jerusalem and rebuild the fortifications, in violation of the treaty, but they didn't return the territory. More than 11 years after agreeing to the 10 year treaty, in 1240 CE the non-Muslim Tartars conquered Jerusalem from the Christians, which should've been returned to the Muslims the year before. The Christians retook Jerusalem in 1241 CE from the Tartars and kept it until 1244 CE when the Muslim Khwarizmis reconquered it, since by that time it was clear to the Muslim world that crusaders don't honor their treaties no matter how generous the terms are. What happened in response to the Muslims getting Jerusalem back? Why the 7th crusade was launched by Pope Innocent IV in 1248 CE. The 7th crusade managed to capture the city of Jerusalem, but within 48 hours the local Muslim tribesmen defeated the entire crusader army and drove them out, later on capturing and ransoming the Catholic "Saint" King Louis IX of France; who gained his fame for having went on multiple crusades and dying in Tunis as a result of fever during his last crusade. After this surprisingly quick defeat a Templar knight sadly declared, "*It seems that God wishes to support the Turks to our loss*" and it was also said, "*Anyone who wishes to fight the Turks is mad, for <u>Jesus Christ **does not** fight them</u>*". When I had a crusader mentality I believed all these defeats were simply flukes, as a Muslim I think differently. Religion aside, if interpreted objectively the only successful crusade depended heavily on a Muslim traitor. All the others were drastic failures that are difficult to blame military reasons for. The one commonality most all crusades from the 11th-

15th century CE have is their endorsements by the Catholic Popes. Today Catholicism teaches that the popes were and are infallible, meaning never wrong, meaning it's obligatory for Catholics to believe that all popes were free from errors. Yet recent popes have said crusader era popes were wrong; thereby not even believing in their own religious doctrines which they propagate. This makes it impossible to believe in papal infallibility because if today's pope is right to say that earlier popes were wrong that means Catholic doctrine is wrong about infallibility and that today's pope can be wrong too. In a nutshell the modern popes are either intolerant liars or it's wrong to believe in Catholic doctrines; or it's both. Pope Benedict XVI who served as Pope of the Catholic Church from 2005 CE to 2013 CE said this in regards to Islam: "***Islam has a total organization of life that is completely different from ours***; *it embraces simply everything,....There is a very marked subordination of woman to man; there is a very tightly knit criminal law, indeed, **a law regulating all areas of life, that is opposed to our modern ideas about society**. One has to have a clear understanding that **it is not simply a denomination that can be included in the free realm of a pluralistic society**.*"

Few today wage jihad correctly and that is why Muslims are not supreme in the land, it's because Muslims aren't fighting Islamically that they have not been victorious. Once Muslims start practicing Islam correctly and waging jihad correctly according to the rules of Islam, then Allah will give the Muslims victory. The reason Muslims do not run the world today is because they don't practice Islam or wage jihad correctly. Plus before you can fully practice Islam correctly you have to make Hijra. Practicing Islam correctly is a prerequisite to waging jihad correctly which is a prerequisite for Muslim victory and the solution to all of humanity's problems. Victory is determined based on how well Muslims are practicing Islam, nothing else. That's why the enemies of Islam are so afraid of Islamic Fundamentalism and misleadingly depict the fundamentalists as extremists. The only reason Muslims have ever been able to win any military engagement is because of their faith. At nearly every time in history disbelievers have always been more numerous than the Muslims with more military capabilities and ferocity. The Muslim generals who conquered Persia, Sham, Africa and Andalus always told their soldiers that if the Muslims sinned more than their enemy then they will be defeated. Military history proves their statements to be true. In comparison the less disbelievers practice their false religions then the more their military prowess increases. Military weapons don't make a difference when Muslims wage Jihad, the Islamic military history proves this. Unfortunately many Muslims don't know that the weapons a Muslim wages Jihad with is Taqwa, Iman and Dua. Especially in Palestine today. For instance Muslim kids in Palestine use slings to throw rocks at Israeli tanks and soldiers when they come to kill Muslims. Now it's not forbidden for Muslims to use weapons to fight and it's not forbidden to use modern weapons either. In fact Muslims lost Russia because some foolish scholar told Muslims it was sinful to defend themselves with guns and told them to defend themselves with spears instead. Yet there's a sahih hadith narrated from the prophet Muhammad collected in Al-Adab Al-Mufrad by Imam Bukhari # 905 which says what in english means:

'Abdullah ibn Mughaffal al-Muzani said, "The Messenger of Allah, may Allah bless him and grant him peace, forbade slings. He said, 'They do not kill game nor injure the enemy. They gouge the eye and break the teeth.'

So it'd be better for Muslims in Palestine to throw stones with their hands than to use slings if they did so desiring to follow Muhammad pbuh. People might think I'm crazy and unrealistic, but truly those stones thrown by hand would do more damage to those tanks than those thrown by slings. Because when a true Muslim throws, it's not them who throws but it is Allah who throws. Whereas Western military powers fear Islamic fundamentalists' rhetoric because they know the truth. Any rational person would think such advice is crazy and benefits Israel, yet America and Israel loathe such advice because they know firsthand it's true. They'd rather Muslims be using slings than follow the Sunnah. The enemies of Islam and the Muslims don't fight us in order to kill us, they fight us to make us abandon Islam and disbelieve. It's a matter of religion not military. The victory will only come from Allah not the AK, or the airplane, or any atomic bombs. The only ally a Muslim ever needs to win is Allah, without Allah there is no victory but only disgrace, even if people claim otherwise. Fighting according to the Sunnah would also mean using other weapons, like dua. Of all weapons, dua is the most dangerous, successful, safe and useable by all Muslims at all times. Which is why one should always make dua for the Mujahideen of the present and future.

Another aspect of Jihad which many, even Muslims, neglect to report on is women's rights. What do I mean by women's rights and Jihad? Well amongst women there are mothers, wives, sisters, daughters, aunts and such. These women have rights according to the laws of Jihad. However I'm not talking about the rights of female soldiers or civilian non-combatants, I'm talking about the rights of Muslim women and the female relatives of Muslim Jihadis. These "women's rights" are not given to women by the unislamic so-called "civilized modern nations", especially in the West. Yes I am saying that women in the West are being oppressed by the very militaries who send men into battle allegedly off fighting on their behalf. Now let me explain. Women have sexual desires, some of them might not like men to know this, but they do have sexual desires. Therefore when a wife's husband is in a military and he gets deployed to fight, that means the wife will not be able to have sex with her husband while he is away. Islam has mandated that a wife has sexual rights over her husband, and it goes both ways. For women these rights include a wife's right to have sex at least once every 4 months if she so desires. While if a man were to refuse to have sex with his wife for over 4 months, she can legally use that as a reason to get divorced in a Shariah court. Since women have this sexual right then the rules of Jihad have to honor this right of women. Therefore during the Khilafah of Umar bin Khattab when Muslim men were fighting in distant places away from their families for extended periods, the military had rules implemented to accommodate women and their rights. For instance women have a right to not labor for money and their husband or male relatives must provide for her. Yet if they are off fighting a war then who is going to take care of the woman? Well since the Islamic government is the one who deployed their husband or male provider then the government must pay for all the needs of the woman until her man gets back to take care of her. She is not expected to use her own wealth, the government must take care of her. Obviously the western governments do not take care of women like this, they simply don't care and won't take care of them while their men are off fighting in war. They tell women go work for yourself or die, and that's why women entered the western workforce during WWI and WWII, because the governments were letting them die and refused to take care of them. Furthermore the time a man can legally spend fighting away from his wife and family is also regulated to be a maximum of 3-4 months, because a wife has the right to have sex and daughters have rights to see their father, mothers have rights to see their sons, sisters have rights to see their brothers and all women have rights in Islam which nobody can take away from them, even if it's for the sake of Allah and the Muslim nation through fighting Jihad. One individual woman's right is more important in Islam than the government or the military. Thus any soldier in an Islamic army can never be deployed for more than 4 months at a time, unless they get special permission or have special circumstances such as no female relatives. Soldiers legally have to come back and spend time with their wife unless their wife voluntarily gives them permission not to, if the wife wants her husband back within 4 months and he chooses not to return then she can divorce him or sue the government for not sending her husband back. However some women might not be able to go 4 months without their husband and if so they can request their husband be returned earlier. One opinion, though weaker, is that the reason a Muslim man is allowed no more than 4 wives maximum is because a women is entitled to have sex with her husband once every 4 days if she so desires and the husband would be obligated to comply. Although I must stress this is a weaker opinion regarding the length of time a wife who wants sex with her husband can Islamically be deprived, I just thought you would be interested to know some Muslim scholars said that Muslim wives have the "right to have sexual relations with her husband every 4 days" if she so desires; Imam Abu Talib al-Makki even added "*if he knows she needs more, he is obliged to comply*". Don't get too excited. That's just something a shy wife with sexual desires might want to inform her husband of when she thinks he needs a reminder. Also women in Islam have the "right to a chaperone", meaning any woman who wants to travel about has the right to have her husband or a male relative come with her to accompany her for safety and companionship reasons. If a women wanted to go somewhere and her husband refused to accompany her and/or her father/brothers/uncles refused to accompany her then she could report them to the state and they'd all get in trouble for oppressing her by not traveling with her. Muhammad pbuh even forbid women from traveling a day and night's length journey without her husband or a male relative because women have this right to protection and companionship. Some in the West confuse this to think women can't go out under Islam, but this is completely false and Muhammad pbuh even forbid men from forbidding women from going to the masjid. It's simply that women have such a high status in Islam that men have to protect them when they travel and must help them get where they want to go, carrying their stuff, driving, etc. For women to have to take care of themselves and travel alone when they go out is considered to be oppressing them in Islam. Women get lonely and scared so they all have a right to protection and companionship. Now unfortunately some sinful Muslim women actually abuse this right of theirs and drag their husbands to sinful places like casinoes or pubs, but this is oppression of the husband because they are not supposed to be going to sinful places and thus husbands/male relatives can and should forbid women from going to sinful places. So that's where the whole Islam doesn't let women leave the house nonsense comes from, it's because the men don't accompany their women to sinful places. Thus when the women leave on sinful outings by themselves, the Muslim government then sees a woman all alone who could be lonely or harmed so they ask her if she has a husband or family and why aren't the men in her life accompanying her as is her right. The government asks the Muslim women this because the men are in legal trouble because of not accompanying her when she travels, but when the Muslim government finds out it's because she is going to a sinful place then they side with the man and get female officials to escort her back home because she shouldn't have been going to a sinful place anyways. Thus the whole islamic society is facilitated to serve women unless they are doing something sinful, then it stops them. Basically in Islam if a woman is not sinning then her husbands and male relatives are her servants, but if she is sinning then she will get in trouble with the state and told to obey her husband/family when they tell her not to do sinful stuff. Thus the special rights women are given under Islam are only alloted to her when she is in the right, this way peoples' rights cannot be abused unless they are in the West/unislamic society, or the Muslim society becomes ignorant of the rights of women. Yet still, in Islam every woman has a right to have her husband or male relatives be her chaperone everywhere she goes. Some stupid guys might think this is a type of surveillance and control but I've talked to lots of non-Muslim women about this and all of them say they love this idea and "wish they could have their husband come with them everywhere they go". Well in Islam that "wish" of non-Muslim women is their "right" and it's been their legal right for thousands of years. To illustrate how highly this right of a woman is held in Islam a Saheeh hadith narrated by Ibn Abbas in Sahih Bukhari relates how one time Muhammad pbuh stated, "*It is not permissible for a man to be alone with a woman(they aren't married or related to), and no lady should travel except with a Mahram(husband or relative)* "after which a Muslim got up and said "*O Allah's Messenger! I have been enlisted in the army for such and such battle and my wife is proceeding for Hajj.*" So now what do you think happened? The man was a Muslim soldier about to wage Jihad in battle but his wife was going to be doing her obligatory pilgrimage to Hajj all by herself, yet Muhammad pbuh said women shouldn't travel alone. So what did Muhammad pbuh say to this Jihadi about to fight in battle? He said, "*Go, and perform the Hajj with your wife.*" Do you know why? Because of the right of the Muslim woman to have someone accompany her while going to do Hajj was greater than the right of the Islamic Muslim army to have one of it's Jihadi soldiers fight in battle. Also keep in mind that in a defensive Jihad things are different so it's not a unbreakable rule for men that they have to accompany their wives instead of wage Jihad, this was an offensive Jihad not a defensive Jihad, every situation has particular rulings and keep in mind the woman was going to Hajj not the shopping center. However while many may then like to take this hadith about how important it is for a man to accompany his wife, keep in mind the hadith also states how men are forbidden to be alone with a woman they aren't married or related to, and guys know why. For the girls who don't know it's because Satan is the third party in such an isolated scenario and adultery and fornication can occur in such moments of isolation where temptation is amplified (in such situations it is extremely likely for mental/emotional adultery/fornication to occur which while not legally punished is still very sinful). Thus such a prohibition nearly eliminates the possibility of adultery and fornication while also giving women the right to protection from being in compromisable positions with men of strange relation or perverted inclinations. This also protects both men and women from being in situations where opportunities and temptations to flirt in private exist. No non-Muslim or unislamic nations have this policy because they religiously do not believe women have these rights. They honestly think that women do not have a "right" to be free from isolated encounters with men they aren't married or related to or that men don't have a "right" to be free from such encounters. Allah says they do and Shariah law gives both men and women their due rights which none can take from them for any reason whatsoever. Under Islamic Shariah men/women can actually say "*No. I don't want to be alone with you and will not be in an isolated situation with you. I as a man/woman have the legal right from God protecting me from being alone with a non-related member of the opposite gender."* Now this prevention of intergender isolation doesn't have to be personal but I am glad God gives this right and hope to live in an environment where it is a punishable crime if someone of the opposite gender violates this right. Really there are so many awkward one on one encounters I'm sure everyone at one time or another wished they were not alone with a member of the opposite gender and there was some social standard that prevented such an encounter from taking place. Islam provides that social standard and Shariah legally enforces it preserving these rights protecting the best interests of mankind. If/when someone tries to take these rights away or violate them, all the Muslims in the world will wage Jihad to defend women's rights or men's rights on their behalf. The 4 month sex rule also applies to the wives of criminals, if a man commits a crime and is in jail, his wife still has a right to have sex with him. She cannot have her sexual rights taken away. Henceforth the religion of Islam invented "conjugal visits". Prior to the Shariah of Muhammad pbuh no conjugal visits were allowed, they just threw you in a dungeon to rot. Yet today many nations don't allow conjugal visits because they don't believe women have these rights and they don't give them their God given rights. In America currently only 4 out of 50 states(8%) permit conjugal visits. Yet the U.S. government claims to be a champion of women's rights? How can that be when 46/50 or 92% of states in America forbid women to have sex with their husbands if either of them are in prison? America has oppressed women and trampled their rights Islam gave them. However that is just the tip of the iceberg of rights which Islam gives women that they don't get from anywhere else. Regarding inheritance of wealth, there are 8 categories of women who inherit but only 4 categories for men. So a woman is twice as likely to inherit wealth in Islam than a man is according to the Shariah laws of inheritance. Meanwhile women shouldn't have financial expenses because the men in their family are religiously obligated to pay for all their expenses. Thus while women need not share any money they inherit, a man is obligated to spend on the women in his family which will cut into his share of any inheritance. Now what do the secular state laws say the rights of men and women are regarding inheritance? Nothing, it's *"Everyone has the freedom to make up their own rules which will lead to injustice and family feuds."* Or it's not quite "freedom to give your money to whatever" because the government taxes the wealth of it's dead citizens via inheritance taxes and gift taxes. So most of these "free countries" are practically governed by grave robbers, since they tax graves/burial expenses, and they are corpse robbers because they tax the inheritances. In comparison Islamic Shariah says God pre-determines which relatives get what when you die and allows one to bequeath only a maximum of 33% of one's estate to non-relatives. The governments get nothing from dead citizens under Shariah. Yet countries proud of "freedom" tax people even after they die thus proving "freedom" means getting oppressed while alive and dead. Oh and keep in mind Islam gave 8 categories of women the right to inherit in the 600s CE. In most of the non-Muslim world at that time women didn't inherit, they were like a piece of property that got inherited by others. Islam gave women inheritance, Kufr gave women as

88

inheritance. Another legal right pertaining to sex which all women have in Islam is the "right to orgasm". In Islam every wife has a right to receive peak sexual pleasure from their husband every time they have sex. Although ladies should note that applies to both spouses and is a mutual right. If the husband routinely does not fulfill this right of his wife and she is consistently dissatisfied with their sexual encounters, or finds them unpleasurable, then the husband is considered to be a transgressing oppressor who is nearly a criminal under Islamic Shariah. Seriously, the religion of Islam teaches that it is abusive torture for a wife to not be sexually satisfied by her husband, but again that goes both ways ladies, for some not having sex can be considered torturous abuse. As a sidenote to guys, it's reported that women need at least 15-30 minutes of caressing before being able to climax; selfish sex is sinful. Because of this and the tendency for men to exhaust their sexual desire/ability before they satisfy their wife's desire, this makes marital sex a form of Jihad. Yes, a husband and wife having sex is a type of Jihad. Now if they die in the act they aren't necessarily considered martyrs, it would depend on how they died and other things, but nevertheless the act of marital sex itself is still a type of Jihad which has strict rules and rights for all participants. And if you don't think it's Jihad, then you ain't doing it right. Whereas this right to sexual pleasure from their husbands is just another right women have in Islam that they don't get outside of Islam. Christianity, Judaism, Atheism, Democracy, Freedom, Feminism, nothing gives wives a "right to sexual pleasure from their husband" except for Islam. Even after the "Sexual Revolution" in the West, the women there still don't have this right which Allah gave Muslim women thousands of years ago. Thus it's so ironic and hypocritical when Western nations and soldiers claim to be fighting Muslims for "Women's rights". Non-Muslims don't even have the slightest clue about what "Women's rights" are. Yet Islam does because Allah decreed these rights for those women whom he created. Women in Islam are even entitled to special legal treatment whenever they get their monthly menstruation cycle. They get religious exemptions from certain duties, entitlement to kindness from their husbands/kids and a wife cannot be divorced when menstruating. Also a divorce take 3 months, during that time the husband and wife still live together with him paying for everything and if they have sex during that 3 month time frame when contemplating divorce then the process is cancelled. If they still want to divorce after making up via sex they'd have to restart the 3 month process all over again and they can't start it during the wife's menstruation cycle. In unislamic lands they don't have such divorce rules that make it hard to get divorced, they make it easy and even offer financial incentives. Don't all women deserve the right to be protected from divorce while menstruating? Don't women deserve special legal treatment when they menstruate? Well Islam says women do have such legal rights. Thus truly consider who should be liberating who? Which type of woman is really oppressed? The Muslim women or the non-Muslim women? The non-Muslim soldiers who fight Muslims and/or Islam aren't protecting their women, they are abandoning, abusing and neglecting their women. Just ask the wives of those soldiers, they feel lonesome and the parents of those soldiers want them to return too. So most non-Muslim soldiers are disobeying their parents to wage war. Muslims need permission from their parents to fight in Jihad. That's one of the rights of parents, which Islamic Shariah has given them, that non-Muslim governments usurp and steal. If an Islamic government were waging Jihad and a Muslim soldier's Muslim parents(yes they have to be Muslim) did not want them to be gone, or fighting, then they would simply tell the government and the soldier would be returned to their parents. Also if a Muslim soldier dies in battle then all the spoils that he earned through his efforts are given to his family. That has been the Islamic policy since the time of Muhammad pbuh, to give the spoils won by a soldier to their inheritors after they die, no other religions teach this nor did non-Muslim nations offer any financial compensation at all to the relatives of killed soldiers until only very recently. Furthermore no Muslim man would ever put his nation or ideology over his wife, kids or relatives, he knows his duty to them takes precedent and a Muslim man would never ever leave their family to fight for any worldly reason. Not even for money, friends or glory. Muslim men only fight when the rights of God or the religion of Islam requires them to do so. Fighting is never a hobby or a profession. If being a soldier is someone's profession then such a person is mentally insane and wicked since it is their job to wage war. Simply by having their soldiers away for so long and not paying for the total maintenance of the womenfolk and families of those soldiers while the soldiers aren't at home the non-Muslim governments are oppressing the people in their own country, when trying to bring their ways of oppression to the Muslim world. That truly is what they do, they bring corrupting oppression to the Muslim world and take away rights but call it "freedom" or "liberation" or "modernization" or "civilization". Overall there is by and large two main "rights" which Islam does not give women for which the non-Muslims claim to be waging war in order to provide. Islam does not permit women to dress any way they please in public or say whatever they want, because female immodesty causes so many problems and corrupts society as well as women themselves and a women's tongue can be a weapon of mass destruction. The fastest way to destroy a nation is to give everyone permission to say or write whatever the hell they want. One of the next fastest ways to destroy a nation is to let the women dress immodestly. Yet men in Islam aren't given the rights to dress however they want either, they must keep their shirts on, and also cannot spew filth from their lips, such as the type which the Western world calls "flirting". Although if we are honest, can we really say society is better off without a modest public dress code and "freedom of consequences of speech"? Because basically that's all a non-Muslim unislamic society can offer women/people, a sinful non-existent dress code and freedom to use their tongue in public and private for evil. Yet in such a system women don't even get that, because if they aren't dressed sexy for perverts non-Muslim men don't care what she is saying, and even if a few do their words lose value because of having to compete with and use sex appeal for "respect" or attention. So for women "freedom" practically just means forced public nudity. If they take off their clothes then they can, in theory, speak freely, but get ignored/disrespected. Whereas under Islam women keep their clothes on, say only good things, get respected, served and nurtured. As an example of this "fight for rights" in 2016 CE when the soldiers of Western nations and the Kurds took over villages in Syria near Manbij they recorded a video of what their "liberation" entailed. It showed a woman who once "liberated" took off her clothes and threw them into the air while she joyfully shouted "Freedom!" Seriously that's what happened. The soldiers who "liberated" her were even more pleased than she was, they really enjoyed and praised her "love for and expression of freedom". To them it made all their efforts seem worth it, to bring freedom/nudity to Muslim women. The Western "liberation" and "freedom" literally means taking off women's clothes and that is what happens. Yet they rarely ever report this stuff in the western news outlets, and they don't show the videos because the western taxpayers would be outraged if they knew their soldiers were literally fighting in order to get women's clothes off. Instead they say they are freeing Muslim women from oppression, but it's a major sin for Muslim women to rip off their clothes or dress the way the "liberated" women in the middle east do. So it makes one wonder what do they do with all the Muslim women who keep their clothes on and refuse to remove their niqab, burqa or hijab? Well you don't have to wonder, because it's well known what happens to such women. They get raped again and again and again by many different males under the excuse they are *"witholding information"*, so US, UN and NATO soldiers simply employ rape as a form of "enhanced interrogation techniques". Unfortunately it doesn't stop there but extends to the Muslim women being forced on a daily basis to drink the sperm of their captors/interrogators/rapists instead of water. Documentation proves this is what the US Marines did in Abu Ghraib prison in Iraq, for years and there is evidence proving that this is what many "freedom fighters" in Muslim countries do. Yet these are just a few of the many reasons why Muslims wage Jihad in response. But remember there are rules to Jihad, Allah made those rules and they must be followed since Allah punishes any and all who break the rules. Thus it takes a very special type of character to correctly wage Jihad, motive and physicality is not enough, one needs knowledge, sincerity and immense spiritual fortitude. If the rules of Jihad aren't obeyed then it can lead to hell, but if they are followed then the rewards are immense and in paradise one gets whatever one wants and nobody can say nothing about it. God made paradise and God makes the rules, if God says those who go to paradise get whatever they want then that's what it is. So if say some guy is killed waging Jihad, goes to paradise and wanted 72 virgin wives in paradise, then he would get that. Yet if his earthly wife joined him in paradise she wouldn't feel any sore feelings because it's paradise and it's impossible to have negative feelings there. Women get whatever they want in paradise too and their husband won't care. One must remember that paradise is paradise for everyone, so if say a guy wanted to have a 50 year long kiss in paradise but his earthly wife didn't want to have a 50 year long kiss then it would be impossible for both to get what they wanted at the same time. Thus God makes it such that everyone can get everything they want without anyone getting upset or having another person's pleasure interfere with their own happiness in any way. Plus one must remember that people will still have private parts in paradise. Meaning men in paradise will still have penises and every guy knows that if they had to spend eternity without any sexual activity they would not call such a place paradise. Mr. Penis would not be happy and in paradise every body part is happy. It might be crude but that's the truth, even if men don't want to publicly admit it. God knows what women want and what men want even if they may not admit it or even know what they really want. What God gives people in paradise is greater than their imagination. That's why the test of life is hard and not easy, because paradise is so pleasurable it requires hard work, but even then our hard work isn't enough and paradise is only a form of charity from God to those who believe whom he forgives for their sins and is pleased with by their good deeds. Thus the blessings of paradise can never be compared to this world or equated with what is "fair", but the same also applies to the hellfire and it's eternal torments which we don't want. In reality nobody has the right to criticize what another gets in paradise or what another wants in paradise, if you created paradise then you can make the rules. Since God made paradise then God alone determines what people in paradise get, the same applies to hell. Life consists of two different Jihads. There is our Jihad to worship God and obey and there is Satan's Jihad to get us to disbelieve and sin. We get rewarded if our Jihad is successful while Satan's target (you) will get punished if Satan's Jihad is successful. So there is your Jihad vs. Satan's Jihad. The war began before you were born and will continue until the millisecond you die. Your entire life until your death is a religious war between you vs. your desires + satan and his army. You need Allah's help to win.

Thus far the jihad which includes fighting has primarily been discussed, but that is only one form of jihad; which most people never reach the level to do it properly. With all the rules and restrictions it is actually very difficult to fight for the sake of Allah according to the rules of Allah, let alone to face the difficulties of war itself. Before one could even consider themselves eligible for jihad of the fighting variety they have to do jihad in all other aspects of their life. For example no one can fight leaving their wealth, family, home and risk potentially losing their life if they cannot fight their desires. If they can't control their appetite fasting during Ramadan and fight the urge to eat gluttonously, or eat prohibited things, how can they fight a physical enemy? If they can't fight their pillow to wake up in the morning to pray how can they wake up in the morning to fight an enemy? If they cannot fight Satan's whispers in order to control their tongue from speaking evil then how can they fight the forces of Satan and control their gun to prevent it from shooting evil? If a Muslim man can't remain steadfast and chaste to achieve a total victory over Mr. Penis how will they achieve a victory against the leaders of disbelief? If they cannot control their spending habits from purchasing sinful or wasteful things and instead spend hard-earned lawful wealth in charity, then how can they ever spend for the sake of Allah in the cause of jihad? If they can't live as a Muslim off the battlefield how can they ever think of gaining Islamic Martyrdom on the battlefield? To similar effect the Muslim Scholar Imam al-Ghazali said: *"Declare your jihad on 13 enemies you cannot see egoism, arrogance, conceit, selfishness, greed, lust, intolerance, anger, lying, cheating, gossiping and slandering. If you can master and*

96

destroy them, then you will be ready to fight the enemy you can see." If you notice he only listed 12 unseen enemies. Can you guess what the 13th is?

Most jihad has nothing to do with violence. Remember jihad linguistically means "to struggle", when you think carefully about it our entire life is nothing but jihad. We struggle to be polite and respectful to those who insult us. We struggle with sexual temptations that surround us. We struggle to feed our families and raise our children properly. In fact for a Muslim woman enduring pregnancy, giving birth and breastfeeding are types of Jihad that can result in enormous rewards. It may even be harder to give birth than it is to fight on a battlefield, and it doesn't get much easier after the baby comes out either. No matter what type of torture mankind invents, they've yet to come close to matching the pain experienced when a mother gives birth to a child. In fact I struggle to see how mothers can even sin after they've given birth, the hellfire is more painful than giving birth yet so many mothers sin without considering that every sin they commit, like gossip or backbiting, can cause them more pain than childbirth and that will last forever. Seriously if women think giving birth is painful, that's actually physical pleasure compared to just hearing the sound of hell. However giving birth to a child is the easiest part when raising a child to become someone God loves. With the exception of Adam pbuh and Eve, every friend of God was born as a child and had to be raised before they could become a great man or a great woman. Any mother can give birth, but few mothers will do the Jihad to raise the creature they gave birth to well so they become a

friend of God; and even fewer fathers will do their part in
that Jihad. Marriage is also a huge struggle which Muslims
consider to be jihad; but don't get the wrong idea. Islam
prohibits married people from physically abusing each other
and it is a grave sin to harm one's spouse physically,
mentally, financially or emotionally. Learning Arabic is
jihad, in which if it is hard to learn one gets twice the reward
as someone who finds it easy. Giving charity is jihad. Doing
good deeds is jihad and avoiding bad deeds is jihad.
Smiling is jihad, seriously Muhammad pbuh taught Muslims
that smiling is charity but sometimes it is a jihad to smile.
Fasting during Ramadan is jihad. If one is tempted to sin at
night, going to sleep instead would be jihad. Dressing
modestly as Allah likes is jihad. A man allowing their beard
to grow is jihad. While a woman wearing niqab or hijab is
jihad, especially in modern times. Being kind and obedient
to your parents is jihad. For men praying 5 times a day in
the masjid is Jihad and for women sometimes trying to get
the men in their family to pray 5 times a day in the masjid is
Jihad. To forgive someone who wronged you is Jihad, it
really is; it's much easier to fight than to forgive. When
wrong to admit you are wrong saying "I was wrong." is the
hardest thing to say to someone. Note that statement "I was
wrong." has a period, there are no "buts" or "ands" following
or added to that ever difficult statement which most of us
would rather die than utter. Although it is sometimes even
harder to say "I forgive you." when someone else is wrong or
wrongs you, especially if they don't think they are wrong or
ask for mercy. Truly saying "I forgive you." is much more
important and pleasing to God than telling someone "I love
you." Every time I've ever been in a situation where I've

been wronged and for personal reasons didn't want to forgive another person, I've regretted it and wished I forgave them sooner. Every time I delayed forgiving someone I've regretted it though I always felt I never would. Don't make the same mistake. The world would be a much better place if people said "I forgive you." more than they said "I love you." Most of us have diarrhea of the ego which makes us say anything but "I was wrong.", "Sorry, please forgive me." or "I forgive you." Yet frequently such words can be words of Jihad. Writing this book is jihad which I hope Allah accepts. That is the key to jihad, one must do it appropriately the Islamic way with sincerity in order for it to be accepted and get rewarded for it. If I have written something wrong, in the wrong way, or for any reason other than pleasing Allah in here then this book could be a reason that causes me to burn in hell. You might think this book is good, bad or both, but your opinion is virtually worthless the only thing that matters is whether Allah is pleased with it. Even if this book is credited to me as a good deed it may be that later on I do a bad deed which could erase the good deed, putting me in the negative. None of us has any guarantee that any of our deeds are accepted, and none of us has any guarantee that our sins haven't erased any good deeds we may have done. It is only when we die that we will have an indication of what Allah thinks of us. Because one could do jihad in many ways but if it's not done solely for the sake of Allah according to his instructions, then it's being done for someone or something else and Allah will not reward one for something that wasn't done for his sake only. Therefore as long as we are alive we must strive to do our best to please our Creator, in Arabic the phrase for that is "Al-jihad, fisabi-lil-lah".

Notwithstanding the unjust way Muslims have been treated, that does not give any Muslim the right to do injustice to non-Muslims, no matter how bad the treatment is no matter for how long. Unjust actions are still unjust and sinful, even if it is a reaction to injustice and oppression. This is something many fail to understand Muslim and non-Muslim alike, but this is what Islam teaches. Sometimes malicious war crimes are committed, even though self-defense is allowed the victims justly retaliating can easily be led to extremes by Satan and commit war crimes themselves, making the Good vs. Evil scenario turn into Evil vs. Evil. This only adds fuel to the fire which will result in all involved being added as fuel to the hellfire, despite one side being good and correct in the beginning. It is better to forgive. Those quickest to forgive will be the quickest to enter paradise and will be shaded on the Day of Judgment when the sun is a mile away and people are drowning in sweat. However while it is better to forgive, sometimes forgiveness will exasperate the problem and cause it to continue if not worsen. In such a case everyone who retaliates must be conscious of not exceeding the limits of justice because this is exactly what Satan wants to happen as a result. If the bounds are overstepped and the oppressors are oppressed, then they will seek revenge and feel justified oppressing again and the cycle of oppression will continue to continue as it has throughout human history. Patience and Forgiveness are important to remember as well as self-restraint. The best example of this is Muhammad pbuh.

Muhammad pbuh was a merchant by trade and was
known as the most trustworthy in the community before his
prophethood began at the age of 40. Since he was so upright
and honest the inhabitants of Mecca would give him their
property for safekeeping in storage, since they knew he
would take care of it for them. When he began preaching
many of his family members and neighbors rejected and
slandered him, turning away from the message of Islamic
monotheism. They would treat him in the worst of ways.
His neighbor Umm Jamil bint Harb bin Umayyah would put
thorns on the Prophet's doorstep pbuh every night intending
to injure him. Blood, urine, feces, animal intestines and all
manners of filth would be thrown at him and his house.
One time an incident occurred where Muhammad pbuh was
praying and while in prostration with his forehead on the
ground one of his own relatives put camel intestines on his
back. Muhammad pbuh remained in prostration praying
until his daughter Fatima came and took the entrails off of
his back. This is just one of the many abuses Muhammad
pbuh endured from his relatives and enemies. As time
passed the persecution got worse and worse until an
embargo was decreed on all Muslims. For 3 years the
pagans of Mecca boycotted Muhammad pbuh and all who
believed in Islam in every way possible so that no Muslims
were allowed to buy/sell anything at all from anyone in
Mecca, or get any work, or marry anyone in Mecca. The
pagans would even meet foreigner traders and buy
everything they had just so the visiting traders had nothing
left to sell to Muslims. The Muslims were forced to live on
the outskirts of the city in the desert where after the
cumulative savings of the Muslims had been exhausted they

survived by eating the leaves of trees and some would even
eat their leather clothing. Have you ever been so hungry
you ate your only pair of shoes? This is what the Muslims
experienced, and what made it worse was that this boycott
was initiated by the non-Muslim relatives of the Muslims in
a tribalistic society. In tribal societies families were a lot
closer than families are today but still due to practicing
Islam the Muslims nearly starved to death. Nonetheless not
a single Muslim apostated during the 3 year embargo.
Eventually this embargo became so severe pagans felt it
would be a crime against humanity to continue the boycott
so they debated amongst themselves and ended the boycott
allowing the Muslims to interact and transact once again in
Mecca. However Islam was still persecuted and the pagan
violence on Muslims increased. Without exception every
single person who was a Muslim in Mecca was physically
harmed because of it. In the year 620 CE, ten years after
Muhammad pbuh became a prophet, 6 idolaters from
Medinah came on a pilgrimage to Mecca, so Muhammad
pbuh preached Islam to them and they became Muslim.
Next year in 621 CE, 12 Muslim men from Medinah came
and pledged to practice Islam , so Muhammad pbuh sent 2
Muslims from Mecca with them to teach and preach Islam in
Medinah. The following year, 622 CE, 73 Muslim men and 2
Muslim women came to visit Muhammad pbuh in Mecca
and took a pledge to defend him militarily and invited him
to come live in Medinah away from persecution where he
could serve as an arbitrator between feuding tribes bringing
Islamic justice to their city. After this second pledge,
Muhammad pbuh instructed the Muslims in Mecca to start
moving to Medinah. Although contrary to what one might

expect, the Meccans didn't want the Muslims to leave because then Islam may spread further and threaten their idolatrous tourism industry. The pagan Meccans would break up families using their tribal authority to forbid the Muslim women and children from emigrating with the men; until weeks of pitiful sorrow made even the cruel pagan hearts relent and allow the Muslim wives and kids to join their men in Medinah. Regarding the Muslim men who tried to leave, the Meccans would hunt them down and kidnap or imprison them since they didn't want any money leaving Mecca and going to Medinah, nor did they want the "Poison of Islam" to spread with the "Radical Muslims" being free to preach their hatred of long-held popular and sacred polytheistic tribal values. Eventually the abuses in Mecca progressed in severity and it was no longer safe for Muhammad pbuh to remain in his hometown of Mecca because a plot was hatched to murder him, 2 months after the invitation to emigrate by the Muslims from Medinah. The Meccan polytheists' plan was that a member of every tribe would play a role in the murder, that way the Muslims wouldn't be able to retaliate since to do so would mean waging war against all the tribes simultaneously. This is the same tactic enemies of Muslims use today. One morning, with his would be assassins waiting outside of his house to slay him, Muhammad pbuh emigrated to the city of Medinah with his friend Abu Bakr, leaving his cousin Ali to stay in his bed so that his enemies wouldn't know he had gone. Ali emigrated to Medinah 3 days after Muhammad pbuh. During those 3 days Ali had been instructed to return all the property that had been entrusted to Muhammad pbuh since he was moving and would no longer be able to hold it for safekeeping. This is something unthinkable today. Imagine you are getting cursed out, harassed, abused and humiliated on a daily basis for 13 years to the point where your tormentors plot to kill you causing you to leave your home and flee. The whole time you are in possession of your enemies' property safekeeping it for them. Despite all the horrendous abuse Muhammad pbuh suffered, not once did he damage or take from the valuables which were entrusted to him. Even during the boycott, Muhammad pbuh would rather starve than wrongfully appropriate the things his enemies entrusted to him for storage. During all the years of persecution Muhammad's enemies didn't even ask him for their property back because they knew he was such a trustworthy man who would not break his promises. They knew no matter how badly they treated Muhammad pbuh or Muslims, that Muhammad pbuh would never "turn the other cheek" becoming two-faced and betray their storage arrangements. Muhammad pbuh was trustworthy and when hit in the face he didn't turn away, he remained the same calling people to worship God alone condemning kufr and kafirs. Muhammad pbuh didn't "turn the other cheek", he'd let his enemies metaphorically hit his cheek again and again without striking back because God did not yet grant permission for him to use violence against people even in self-defense. I'm not saying Muslims today are supposed to tolerate all types of abuse but at that time in history, rather than "turn the other cheek" as Christians like to preach Muslims didn't even flinch. You see there is a big difference between "turning the other cheek" and not even flinching when persecuted, Christians might turn the other cheek once or twice in anger or resentment but Muslims take

the blows patiently without even flinching. After all that
hostility, Muhammad pbuh instructed his cousin Ali to
distribute all the property entrusted to him to the rightful
owners. Muhammad pbuh could've taken it all and ran
instead, but that was not the type of man he was, even
though the pagans had stolen and confiscated the property
of Muslims and plotted to kill him. Regardless of how much
suffering he went through it didn't make him recalcitrant or
unjust. Although the abuse didn't stop when he left Mecca,
because he kept calling people to Islam. In fact as soon as
the Meccans heard Muhammad pbuh left the city, before he
even got to Medinah, they put a bounty of 100 camels on his
head. Simultaneously Ali was giving them their property
which Muhammad pbuh had faithfully kept for them and
ordered to be returned. So while Muhammad pbuh was
fleeing his hometown and giving his enemies back their
property via a proxy, those same enemies, of whom many
were his blood relatives, were publicly promising to pay a
reward of 100 camels to anyone who brought them the head
of Muhammad pbuh. Even today 100 camels is a lot of
wealth, but back then in that climate it would almost be like
having 100 automobiles that gave you milk, reproduced and
could be eaten. It is reported that when Muhammad pbuh
finally arrived in the city of Medinah there were only 98
Muslim men there. 25 Muslim men were from Mecca and
the other 73 Muslim men were of Medinah, while 10 Muslim
women were from Mecca and 2 Muslim women were from
Medinah. There were other Muslims though, some had
moved to Abyssinia, some lived in the desert as Bedouins,
and some unfortunately were unable or chose not to leave
Mecca. On the journey to Medinah Muhammad pbuh met 2

people who became Muslims and joined him in his journey to Medinah. Anyways when you do the math there were about 112 Muslims in Medinah when Muhammad pbuh first arrived. Some people became Muslim on that very day, but when Muhammad pbuh first set foot in Medinah there were about 112 Muslims in the city. The polytheist Arabs and the Jews in Medinah both outnumbered the Muslims, on their own. Many people forget this, however the fact is that the Muslims were the smallest minority in Medinah when Muhammad pbuh arrived. Yet despite Muslims being "the minority" the people of Medinah agreed to let Muhammad pbuh govern the city as it's leader because they were sick of killing each other through intertribal feuds and wanted the justice that Muslims displayed amongst society. So whether Muslims are in the majority or are "the minority of minorities" has nothing to do with whether Islamic law becomes the law of a certain location. After Muhammad pbuh became a leader of a city what was his first official political act of governance? Did he kill everyone who wasn't a Muslim? No, of course not! He had never read the Christian playbook known as the bible. Instead Muhammad pbuh had a city charter drafted that would be the template for governance and reference for Muslims and disbelievers regarding their rights and responsibilities now that he was in charge. This was drafted on his behalf, since he couldn't read or write himself, in the year 622 CE. Now keep in mind the revelations of the Quran had not been completed at this time so from a legal Islamic perspective this charter can not be made today since Islamic law was further developed after the prophet pbuh made this charter. It's similar to how Moses pbuh was leader of people in the desert before he was given the complete legal code and religious instructions by God. So think of this charter as comparable to a theoretical charter of Moses pbuh before Moses pbuh received the full law he was to receive. Thus while legally this charter cannot be replicated today, or in the future, it is still useful and relevant to examine from a historical aspect. Although as another disclaimer I must clarify that the actual charter itself has not survived in physical form to the present day. So what follows is the Charter of Madinah as related by the scholar Ibn Ishaq who was born in Madinah in 704 CE and died by 770 CE. Ibn Ishaq was the first to compile a full biography of Muhammad pbuh after collecting the information from the prophet's companions and the students of the companions of Muhammad pbuh who died before he could personally hear from them. So while Ibn Ishaq is not a direct eye-witness to the charter, he was born in the city shortly after it was made and is the closest source to the original charter that we can examine from our modern age. Now maybe an erudite person will be extra strict and say: *"Well since we don't have the original it can't be trusted."* Whereas that's an overgeneralization regarding trustworthiness of information, however for the sake of argument or not arguing we can still justify examining Ibn Ishaq's account for historical reasons. Now religiously I'd agree, with the principle that we can't use Ibn Ishaq's account to make a legal religious decision unless it were proven 100% authentic. However Ibn Ishaq died by the year 770 CE, so this account is what Muslims themselves during the 700s CE claimed that Muhammad pbuh made as the charter of Madinah when he was the leader. Therefore one cannot say that this charter is a modern fiction, because it's not modern at all. Whether the charter clauses mentioned by Ibn Ishaq are identical to the actual is a moot point historically speaking because Ibn Ishaq's account is a primary source of what Muslims believed about it in the 700s CE. Thus we can say with 100% certainty that since the 700s CE many Muslims have believed Muhammad pbuh made the following charter in 622 CE when appointed as ruler of Madinah. Which again remember even if this were 100% verified as the actual charter clauses, legally speaking Islamic governments today cannot replicate this charter entirely because there are new Islamic instructions sent by Allah and commanded by Muhammad pbuh after this charter was made, such as Jizya etc. Yet all that aside I still think it is beneficial to mention what many believe to be the charter of Madinah in 622 CE made by Muhammad pbuh, when as a minority the 112 Muslims established Shariah law in a city of about 2,000 Jewish men and about 4,000 Arab polytheists. Now the number of women and children in Madinah is unknown, the number of men is reasonably known because such numbers were known to most because of military reasons and tribal pride to boast of large numbers of "fighting men". Thus keep the context in mind that there were only about 100 Muslim men and Muhammad pbuh is in charge over a city of about 6,000 disbelieving men. So regardless of Ibn Ishaq's account of this treaty not being the physical charter itself the known "facts on the ground" indicate that clearly Muhammad pbuh had to have some redeeming qualities and justice to be able to rule over a city where the Muslim men, half of whom were foreign refugees, were outnumbered 112 to 6,000. It's safe to say that democracy and "majority vote" was definitely not the reason

Muhammad pbuh was made the leader of Madinah. Also he did not spread Islam "by the sword", most of the Muslims didn't even bring weapons with them to Medinah. I mention this for modern political lessons in that people who claim Muslim minorities, or foreign Muslim minorities, should not rule non-Muslim majority lands simply don't know what they are talking about. Because Muhammad pbuh did it when the Muslim men were only 1.86% of the total population of men in the city of Madinah. Muslims did not vote Muhammad pbuh into leadership. Also about 24.7% of the Muslims were foreigners from Mecca, including Muhammad pbuh. Thus for democratic people today claiming it's not right or possible for Muslim minorities to establish Shariah or rule in non-Muslim majority lands, or have a foreign Muslim lead their nation, they should keep in mind that as a foreign refugee Muhammad pbuh was made the leader of a city and established Shariah peacefully in a place where 98.14% of the people were disbelievers and did not believe in Islam. People today talk about how 99% of global wealth is owned by 1% of the population, well when Muhammad pbuh established Shariah in Medinah the Jews had the money and the idolatrous polytheistic Arabs had the numbers, the Muslims only had the true religion and the help of Allah. Yet the Muslims, without the money and without the numbers established political authority with Shariah law creating an Islamic State in a city where over 98.14% of the people were non-Muslims. Obviously no modern politician today could do this unless they were of great moral character like Muhammad pbuh was. But that's enough of a modern political lesson to take in from this event. That was a real political event that took place in the real world and those non-Muslims were real people who were religious too, moreso than the average person today is. For the modern majority of non-Muslims who think that an Islamic State ruled by Shariah law means "*No non-Muslims live there.*", this notion is completely false. Muhammad pbuh established the first Islamic State in a city which had a non-Muslim population of over 98%. Clearly that in itself is a sign of prophethood, because only a prophet of God with the mannerisms of Jesus pbuh and the justice of Moses pbuh or Solomon pbuh could possibly rule over such a large number of people who didn't believe in his religion using his own religion as the basis for the legal system. Therefore although the original document of the charter is not available to study firsthand, I think you would be interested in the historical record of what the Charter of Medinah is believed to have said. The Charter of Medinah as related by Ibn Ishaq which Muhammad pbuh is believed to have made as leader says:

In the name of God the Compassionate, the Merciful.

1. *This is a document from Muhammad the Prophet, governing relations between the Believers i.e. Muslims of Quraysh and Yathrib and those who followed them and worked hard with them. They form one nation - Ummah.*

2. *The Quraysh Muhajireen(Muslim Emigrants/refugees) will continue to pay blood money, according to their present custom.*

3. *In case of war with anybody they will redeem their prisoners with kindness and justice common among*

Believers. *(Not according to pre-Islamic nations where the rich/poor get treated differently).*

4. *The Bani Awf(* tribe of Medinah) *will decide the blood money, within themselves, according to their existing custom.*

5. *In case of war with anybody all parties other than Muslims will redeem their prisoners with kindness and justice according to practice among Believers and not in accordance with ignorant/unjust pre-Islamic notions.*

6. *The Bani Saeeda, the Bani Harith, the Bani Jusham and the Bani Najjar* (other Arab Medinan tribes) *will be governed on the lines of the above (principles)*

7. *The Bani Amr, Bani Awf, Bani Al-Nabeet, and Bani Al-Aws* (other Arab Medinan tribes) *will be governed in the same manner.*

8. *Believers will not fail to redeem their prisoners they will pay blood money on their behalf. It will be a common responsibility of the Ummah and not of the family of the prisoners to pay blood money.*

9. *A Believer will not make the freedman of another Believer as his ally against the wishes of the other Believers.*

10. *The Believers, who fear Allah, will oppose the rebellious elements and those that encourage injustice or sin, or enmity or corruption among Believers.*

11. *If anyone is guilty of any such act all the Believers will oppose him even if he be the son of any one of them.*

12. A Believer will not kill another Believer, for the sake of a disbeliever (non-Muslim). (even though the disbeliever is his close relative).

13. No Believer will help a disbeliever (non-Muslim) against a Believer.

14. Protection (when given) in the Name of Allah will be common. The weakest among Believers may give protection (In the Name of Allah) and it will be binding on all Believers.

15. Believers are all friends to each other to the exclusion of all others.

16. Those Jews who follow the Believers will be helped and will be treated with equality. (Social, legal and economic equality is promised to all loyal citizens of the State, non-Muslims while religiously not equal to Muslims in the sight of God, or Muslims, they will still be treated with legal equality and fairness/kindness).

17. No Jew will be wronged for being a Jew.

18. The enemies of the Jews who follow us will not be helped.

19. The peace of the Believers (of the State of Medinah) cannot be divided. (it is either peace or war for all. It cannot be that a part of the population is at war with the outsiders and a part is at peace).

20. No separate peace will be made by anyone in Madinah when Believers are fighting in the Path of Allah.

21. Conditions of peace and war and the accompanying ease or hardships must be fair and equitable to all citizens alike.

22. *When going out on expeditions a rider must take his fellow member of the Army-share his ride.*

23. *The Believers must avenge the blood of one another when fighting in the Path of Allah.* (This clause was to remind those in front of whom there may be less severe fighting that the cause was common to all. This also meant that although each battle appeared to be a separate entity it was in fact a part of the same religious war, which affected all Muslims equally.)

24. *The Believers (because they fear Allah) are better in showing steadfastness and as a result receive guidance from Allah in this respect. Others must also aspire to come up to the same standard of steadfastness.*

25. *No disbeliever will be permitted to take the property of the Quraysh* (the name of the enemy pagan Meccan tribe) *under his protection. Enemy property must be surrendered to the State.*

26. *No disbeliever will intervene in favour of a Quraysh, (because the Quraysh having declared war are the enemy).*

27. *If any disbeliever kills a Believer, without good cause, he shall be killed in return, unless the next of kin are satisfied (as it creates law and order problems and weakens the defence of the State). All Believers shall be against such a wrong-doer. No Believer will be allowed to shelter such a man.*

28. *When you differ on anything (regarding this Document) the matter shall be referred to Allah and Muhammad (may Allah bless him and grant him peace).*

29. *The Jews will contribute towards the war when fighting alongside the Believers.*

30. *The Jews of Bani Awf will be treated as one community with the Believers. The Jews have their religion. This will also apply to their freedmen. The exception will be those who act unjustly and sinfully. By so doing they wrong themselves and their families.*

31. *The same applies to Jews of Bani Al-Najjar, Bani Al Harith, Bani Saeeda, Bani Jusham, Bani Al Aws, Thaalba, and the Jaffna, (a clan of the Bani Thaalba) and the Bani Al Shutayba.*

32. *Loyalty gives protection against treachery. (loyal people are protected by their friends against treachery. As long as a person remains loyal to the State he is not likely to succumb to the ideas of being treacherous. He protects himself against weakness).*

33. *The freedmen of Thaalba will be afforded the same status as Thaalba themselves. This status is for fair dealings and full justice as a right and equal responsibility for military service.*

34. *Those in alliance with the Jews will be given the same treatment as the Jews.*

35. *No one (no tribe which is party to the Pact) shall go to war except with the permission of Muhammed (may Allah bless him and grant him peace). If any wrong has been done to any person or party it may be avenged.*

36. *Any one who kills another without warning (there being no just cause for it) amounts to his slaying himself and his*

household, unless the killing was done due to a wrong being done to him.

37. *The Jews must bear their own expenses (in War) and the Muslims bear their expenses.*

38. *If anyone attacks anyone who is a party to this Pact the other must come to his help.*

39. *They (parties to this Pact) must seek mutual advice and consultation.*

40. *Loyalty gives protection against treachery. Those who avoid mutual consultation do so because of lack of sincerity and loyalty.*

41. *A man will not be made liable for misdeeds of his ally.*

42. *Anyone (any individual or party) who is wronged must be helped.*

43. *The Jews must pay (for war) with the Muslims.* (this appears to be for when Jews are not taking part in the war. Clause 37 deals with occasions when they are taking part in war).

44. *Yathrib*(name of Medinah before it was renamed Medinah) *will be Sanctuary for the people of this Pact.*

45. *A stranger (individual) who has been given protection (by anyone party to this Pact) will be treated as his host (who has given him protection) while (he is) doing no harm and is not committing any crime. Those given protection but indulging in anti-state activities will be liable to punishment.*

46. *A woman will be given protection only with the consent of her family (Guardian).* (this avoided inter-tribal conflicts).

47. *In case of any dispute or controversy, which may result in trouble the matter must be referred to Allah and Muhammed, The Prophet of Allah will accept anything in this document, which is for (bringing about) piety and goodness.*

48. *Quraysh*(name of the enemy pagan tribe of Mecca) *and their allies will not be given protection.*

49. *The parties to this Pact are bound to help each other in the event of an attack on Yathrib*(Medinah).

50. *If they (the parties to the Pact other than the Muslims) are called upon to make and maintain peace (within the State) they must do so. If a similar demand is made on the Muslims, it must be carried out, except when the Muslims are already engaged in a war in the Path of Allah. (so that no secret ally of the enemy can aid the enemy by calling upon Muslims to end hostilities under this clause).*

51. *Everyone (individual) will have his share (of treatment) in accordance with what party he belongs to. Individuals must benefit or suffer for the good or bad deed of the group they belong to. Without such a rule party affiliations and discipline cannot be maintained.*

52. *The Jews of al-Aws, including their freedmen, have the same standing, as other parties to the Pact, as long as they are loyal to the Pact. Loyalty is a protection against treachery.*

53. *Anyone who acts loyally or otherwise does it for his own good/loss.*

54. *Allah approves this Document.*

55. *This document will not (be employed to) protect one who is unjust or commits a crime (against other parties of the Pact).*

56. *Whether an individual goes out to fight (in accordance with the terms of this Pact) or remains in his home, he will be safe unless he has committed a crime or is a sinner. (i.e. No one will be punished in his individual capacity for not having gone out to fight in accordance with the terms of this Pact).*

57. *Allah is the Protector of the good people and those who fear Allah, and Muhammad is the Messenger of Allah (He guarantees protection for those who are good and fear Allah).*

Clearly that's a very good state charter for 622 CE. At the time such political rights were not granted to anyone else on the planet. Out of all cities on earth at that time this was the best charter being put into effect. Thus one can not just admire, but wonder how this guy in the desert was able to be so just when nobody else on the entire planet had a similar system and no such rights or systems were known to have existed, except in the famous legends of what rule was like in the time of Solomon pbuh. Whether one believes Muhammad pbuh was a prophet or not, everyone must admit that he was a great politician working political miracles. Really the establishment of an Islamic State when the Muslims were hated and persecuted by almost everyone else in Arabia in a city where 98% of the people were non-Muslims is a political miracle. If you think not then just ask around and see what most people say as to whether such a thing could happen in the world today. Soon after Muhammad pbuh was established in Medinah the pagan Meccans went to war with him trying to extinguish the growing Muslim community. They'd make raids on Medinah's agricultural land destroying fruit trees and stealing animals. As a result the Muslims would raid the Meccan trade caravans that bypassed Medinah on the way to Syria. One such raid turned into a battle as the caravan sent word back to Mecca and the Meccans came out with an army to fight and kill the Muslims once and for all. This was called the Battle of Badr and it occurred during the middle of Ramadan in 2 A.H. or March 624 CE. Just 2 short years after the Muslims emigration to Medinah, the battle of Badr consisted of 314 Muslims fighting 950 Pagans. Remember that originally in 622 CE there where only 100 Muslim men in Medinah, although in 624 CE there were 314 Muslim men in Medinah willing to fight for Islam, with other Muslim men who missed the battle since it took place unexpectedly during what was planned to be a routine raid against a trading caravan. Originally the Muslims merely intended to raid a Meccan caravan of 34 people who were returning from selling the goods they confiscated from the Muslims who had emigrated away from Mecca. When the Meccans learned of the attempted raid they sent an army to fight the Muslims, despite the caravan already having escaped safely. Thus the Muslims went from having a 9 to 1 advantage to being outnumbered 3 to 1. Yet despite the odds Muslims won at Badr suffering 14 Muslim casualties, inflicting 70 pagan casualties and taking 70 pagans prisoner. Unfortunately in this battle some Muslims from Mecca who didn't emigrate where forced by the Pagans to come and fight their brethren in faith, sadly some of them were killed in the fray and Allah revealed that they had in reality disbelieved in Islam by not emigrating and supporting the enemies of Islam to fight the Muslims. Fortunately though some of the Muslims forced to fight in the pagan army were taken prisoner, and eventually it was revealed by Allah that those few taken prisoner (since they didn't fight in the battle but merely appeared) were Muslims. Thereupon Allah sent verses of the Quran instructing Muslims who lived in the land of the enemy (which at that time was Mecca) that they must emigrate to the land of the Muslims/Shariah if they are able to, lest they fall into disbelief and die in such a state; due to being forced to help the war effort of the pagans by being in Mecca. Shortly after the results of the Battle of Badr Muhammad pbuh publicly asked the Jews of Medinah to become Muslim and said, "*O Jews, beware lest God bring on you the like of the retribution which he brought on Quraysh. Accept Islam, for you know that I am a prophet sent by God. You will find this in your scriptures and in God's covenant with you.*" The Jewish tribe of Banu Qaynuqa said, "*O Muhammad, you seem to think that we are your people. Do not deceive yourself because you vanquished a contingent of Quraysh having no knowledge of war and got the better of them; for, by God, if we fight you, you will find that we are real men, and that you have not met the like of us.*" Whereas when Muhammad pbuh asked them to become Muslims he wasn't threatening them but just using the event of victory over pagans to tell the Jews they should embrace Islam before they naturally died and went to hell because in such an event they would be akin to the Quraysh some of which recently died as disbelievers. Yet Banu Qaynuqa responded by issuing a threat of war trying to frighten the Muslims and Muhammad pbuh, saying not only would they be willing to fight the Muslims but that they'd win and the Jews issued this threat in the name of God. That was a warning sign of Jewish plots to betray and overthrow the Islamic State. Some time after that warning sign a Jewish man from the tribe of Banu Qaynuqa in Medinah molested a Muslim woman. This Jew was a goldsmith and when the Muslim woman entered his shop to get some jewelry he wanted her to uncover her face and she refused, yet he, like many guys, was a pervert who could not handle being deprived of the sight of a beautiful believing woman's face. So without the Muslim woman knowing he slyly pinned her dress bottom to her seat while she was sitting waiting for the jewelry. When she stood up her clothes came off, leaving her naked. The Jew who did this to her laughed heartily at having declothed a Muslim woman, particularly after she insisted on keeping her face covered to please Allah. The Jew thought it was a funny joke that she refused to uncover her face for him and in response he got her to end up naked in front of many. However a Muslim man who saw what happened was outraged because a Muslim's blood, wealth and honor are sacred(in the sense that they are inviolable and cannot be harmed). This in itself was also a violation of the charter the Jews agreed to abide by as citizens in the Islamic State, they were not supposed to be taking off the clothes of Muslim women shaming them in public then laughing about it. In response the Muslim man reacted by beating up the Jewish man who violated the honor of the Muslim woman. Although he didn't know quite when to stop and the Jewish guy died from the beating. Now the authorities probably should have been informed before this happened, but the Muslim man just reacted spontaneously to the crime he saw. Then rather than the Jews going to Muhammad pbuh to settle the matter, they ganged up and beat the Muslim man to death on the spot. Which again violates the principle of a Muslim's blood, wealth and honor being sacred and was another violation of the charter the Jews agreed to abide by. The Jews legally could not just go killing a Muslim man taking the law into their own hand, and the reason the Jew got killed was because he broke the law to begin with. Also Muhammad pbuh taught that if a Muslim kills a non-Muslim by mistake then they are to be punished but cannot be killed since that would mean non-Muslim lives were equivalent to Muslim lives and would thus imply a believer was equal to a disbeliever. Which is not the case, now if a Muslim intentionally kills a disbeliever then that's different but for a genuinely accidental death of a non-Muslim caused by a Muslim they can't be executed for an accidental death because it was an accident and it would be unjust. But anyways for the record the Jew started it. When the Muslims saw the Jews kill a Muslim man they fought back and killed some of the Jewish killers and then the Jews fought back and killed some Muslims. However this whole incident only involved 1 Jewish tribe, that of Banu Qaynuqa who if you remember prior to this had already threatened to fight the Muslims including Muhammad pbuh. Then the tribe of Banu Qaynuqa all blockaded themselves into a fortress they owned in the southern parts of Medinah and the city was in a state of rebellion. After learning the Jews had killed the Muslim man, Muhammad pbuh decreed that the killing of 1 Muslim was sufficient reason to declare war against the tribe of Banu Qaynuqa. Thus establishing a legal Islamic standard which sadly has not been followed by many "Muslim governments" in the "civilized" modern world. The fortress of Banu Qaynuqa within Medinah, was besieged by Muhammad pbuh and the Muslims. In the fortress there were 700 Jewish soldiers, 300 of which had armor while 400 didn't. This was a real threat to the Muslim state since in the battle of Badr, they had won against the pagans in the month before, the Muslims only had 300 soldiers survive. Yet the Muslims despite being outnumbered and being equipped with less military armaments in both quantity and quality, were firm in besieging the Jewish tribe within their fortress and quelling the rebellion. Why? Because 1 Muslim got killed and that's how an Islamic State/the Muslim world reacts when 1 Muslim gets killed by non-Muslims. You kill one of us and you are at war against all of us and it doesn't matter who it is or where it is. 1 killed Muslim is far too many and is never tolerated under any circumstances unless it was genuinely a freak accident, but in such a case blood money must be paid to the relatives or Muslims if the relatives aren't Muslims; unless the family or the Muslims mercifully forgive the killer and refuse to accept the blood money. Anyways the siege lasted for 15 days until the Jews surrendered unconditionally. Since this was the tribe of Banu Qaynuqa, who were the most ill-mannered towards the Muslims for the last two years, had previously threatened to fight the Muslims and then revealed through their fortification that they had planned this rebellion ahead of time, Muhammad pbuh was not pleased with this betrayal nor in a mood to show mercy. Furthermore the Jews of this tribe were unrepentant and still thought the pervert's "joke" was funny after all that happened. This Jewish tribe was an open hostile enemy of Islam and the Muslims who rebelled within the Muslim state. The Jewish soldiers thereupon had their hands tied behind their back and were to be executed for treason. However a man named Abdullah ibn Ubai ibn Salool physically accosted Muhammad pbuh and refused to let him go unless he spared the lives of the 700 Jewish soldiers who he insisted were his allies. This man would later be revealed by Allah as a hypocrite and enemy of the Muslims. Yet since such revelation had not come and the verses of the Quran forbidding Muslims to be friends or allies with non-Muslims had not yet been revealed, Muhammad relented and ordered the Jewish soldiers be untied. To resolve the matter Muhammad pbuh stipulated that since their tribesman and then their tribe did not follow the agreed upon rules of the charter and fought against the state for 15 days then their tribe would be expelled from the city. Abdullah ibn Ubai ibn Salool again tried to change the prophet's mind but the decree of expulsion was non-negotiable. Hence the Jewish tribe of Banu Qaynuqa were expelled from Medinah in 624 CE and given 3 days to leave, so they went to live in Adhri'at. However the other Jewish tribes still remained in Medinah without any problems. The next major political actions taken by the prophet were assassinations. Muhammad pbuh could handle personal insults but there were some disbelievers who would insult him and advocate war against the Muslims. Thus to protect his honor as a prophet, but mainly to protect the Muslim state from further attacks due to such vehement war preachers inciting war against the Muslims, 3 individuals were targeted for assassination. The first was Asmaan bint Marwaan. Asmaan was an influential tribal leader of Banu Khatamah who slandered the prophet pbuh, called for war against the Muslims and would not allow any of her tribe to become Muslims. Due to her hatred and intolerance of Islam many in her tribe had to keep their Islamic faith a secret to stay alive. Immediately after she was assassinated many people from her tribe openly pronounced their belief in Islam. The second to be assassinated was Abu Ifk Al-Yahudi. He was a Jew who would make hate-filled poems about Islam, the Muslims and Muhammad pbuh which would spread throughout Arabia akin to how pop songs spread throughout the world today. The third to be assassinated was Ka'ab ibn Al-Ishraf. Ka'ab was also a Jewish poet who made slanderous poetry about Muhammad pbuh some of which was even recited by the Quraishi enemies from Mecca. His poetry would make the pagan Meccans cry for sorrow due to their loss at Badr and Ka'ab told them to get revenge and fight the Muslims again until they were wiped off the planet. However Abu Sufyan, the leader of Mecca, was skeptical about this Jewish poet's motives. So he said, "*I ask you by God, which is more beloved to God? Our religion(of polytheistic idolatry) or that of Muhammad and his Companions?*" The Jewish poet replied, "*You are on a more guided path than they are.*" That's how much this Jew hated Islam and Muslims, he was willing to tell a immoral polytheistic idolater, who literally worshipped many statues, that their faith had more guidance than the Islamic faith.

Even though the Jew knew the Islamic faith tells people to believe in and follow all the Jewish prophets and worship only the 1 Creator of the Universe. As a result of Ka'ab's poems and incitement the Meccans agreed to gather another army to attack the Muslims once again. Meanwhile in Mecca Ka'ab continued his poetry attacking Islam, Muhammad, Muslims and even Muslim women. Yet what makes this worse is that this Jewish poet was a citizen of Medinah and had signed the charter drawn up by Muhammad pbuh. So he wasn't just a Jewish enemy, he was a huge traitor in violation of the charter in a plethora of ways who actually arranged for Meccan idolaters to attack his own city. To counter Ka'ab, a Muslim poet named Hassan ibn Thabit denounced with his own poetry whoever hosted Ka'ab in Mecca. Since nobody liked hearing popular poems defaming them for hosting an enemy of Muslims Ka'ab eventually had no place to stay in Mecca. Ka'ab then came back to live in Medinah thinking all would be kosher. Which according to Ka'ab's Jewish faith, his treason was kosher but it was not halal or legal in the opinion of Allah so Ka'ab got assassinated for his crimes. Next year the battle of Uhud took place in 3 A.H. on March 22nd, 625 CE, which involved 700 Muslims fighting 3,000 Pagans who wanted revenge for having lost the battle of Badr the year before; having been inspired by the Jew Ka'ab ibn Al-Ishraf to attack the Muslims again. 1 Christian monk named Abu Aamir ar-Raahib also fought alongside the pagans in this battle against the Muslims. Initially the Muslims had about 1,000 soldiers come to the battle but then a plot by the hypocrite and the previously self-proclaimed ally of the Jews, Adullah ibn Ubai ibn Salool, was put into effect. So Ibn Salool and

300 others left the army with a false excuse claiming they needed to defend the defenseless city. In reality Ibn Salool thought the sudden abandonment of 30% of the army when the Muslims were already outnumbered 3-1 would demoralize them and cause them to lose. Allah soon revealed the hypocrite's real plans and intentions to Muhammad pbuh and the Muslims. Regardless the 700 Muslims prepared to fight the 3,000 revenge driven Meccan pagans. Quickly the pagans lost and retreated, allowing the bulk of the Muslim army to reach their campsite to start looting it while the enemy fled. However 40 out of the 50 Muslim archers (80%), foolishly disobeyed the prophet's orders to stay guarding the rear no matter what happened and left their positions on Mount Ainain(near Mount Uhud) trying to loot the dead enemy thinking the battle was over. During this time the pagan calvary led by Khaalid ibn Waleed attacked in the exposed rear trapping the Muslim army on all sides. Khaalid was a military genius, he and Genghis Khan were the only 2 generals in all of recorded history to never lose a battle. Another specialty of Khaalid was his ability to use a sword with his right and left hand with equal skill, in battle Khaalid wielded two swords at once and rode his war-horse directing it with his legs. After Khaalid's squadron struck the Muslim army in the rear the fleeing pagans saw they could pincer the Muslims so they returned to fight. Muhammad pbuh nearly died in this battle as a result of a blow that hit his head which broke his nose, a tooth and caused 2 chainlinks from his helmet to get embedded in his cheek. The pagan Meccans announced they killed Muhammad pbuh, but although severely wounded he was not dead. However the surprise the

Muslims had when hearing and thinking that Muhammad pbuh had died caused many to simply sit down and stop fighting in the middle of the battle due to the shock of such false news. One moment the Muslims were fighting in Jihad against non-Muslims who were attacking the Muslims in Muslim lands trying to stop Shariah from spreading and then the Muslims stopped waging Jihad, because they thought the prophet was no longer alive. Thus on the battlefield itself they stopped fighting the disbelievers despite the disbelievers continuing to fight and kill Muslims. They mistakenly thought that if the prophet was no longer alive then there was no reason to fight and that as it was obligatory for them to have lived with the prophet under Shariah, with the prophet dead they figured it was no longer obligatory. Hence in a split second because they thought the prophet's time on earth had passed many decided to abandon Jihad and the land of the Muslims thinking only about their own worldly interests and how they could get to safety. Many Muslims fled the battle. In regards to this dangerous unislamic phenomenon where Muslims abandoned Jihad Allah revealed verses in the Quran 3:144-157 which said what means:

*Muhammad is no more than a Messenger, and indeed (many) Messengers have passed away before him. **If he dies or is killed, will you then turn back on your heels (as disbelievers)?** And he who turns back on his heels, not the least harm will he do to Allâh, and Allâh will give reward to those who are grateful. And no person can ever die except by Allâh's Leave and at an appointed term. And whoever desires a reward in (this) world, We shall give him of it; and whoever desires a reward in the Hereafter, We shall give him thereof. And We shall reward the grateful. And many a*

Prophet fought (in Allâh's Cause) and along with them large bands of religious learned men. But they never lost heart for that which did befall them in Allâh's Way, nor did they weaken nor degrade themselves. And Allâh loves As-Sâbirun (the patient). And they said nothing but: "Our Lord! Forgive us our sins and our transgressions (in keeping our duties to You), establish our feet firmly, and give us victory over the disbelieving folk." So Allâh gave them the reward of this world, and the excellent reward of the Hereafter. And Allâh loves Al-Muhsinûn (the good¬doers). **O you who believe! If you obey those who disbelieve, they will send you back on your heels, and you will turn back (from Faith) as losers.** Nay, Allâh is your Maulâ (Patron, Lord, Helper and Protector), and He is the Best of helpers. **We shall cast terror into the hearts of those who disbelieve**, because they joined others in worship with Allâh, for which He had sent no authority; their abode will be the Fire and how evil is the abode of the Zâlimûn (polytheists and wrong¬doers). And Allâh did indeed fulfil His Promise to you when you were killing them (your enemy) with His Permission; until (the moment) you lost your courage and fell to disputing about the order, and disobeyed after He showed you (of the booty) which you love. Among you are some that desire this world and some that desire the Hereafter. Then He made you flee from them (your enemy), that He might test you. But surely, He forgave you, and Allâh is Most Gracious to the believers. (And remember) when you ran away (dreadfully) without even casting a side glance at anyone, and the Messenger (Muhammad) was in your rear calling you back. There did Allâh give you one distress after another by way of requital to teach you not to grieve for that which had escaped you, nor for that which had befallen you. And Allâh is Well¬Aware of all that you do. Then after the distress, He sent down security for you. Slumber overtook a party of you, while another party was thinking about *themselves (as how to save their ownselves, ignoring the others and the Prophet) and thought wrongly of Allâh - the thought of ignorance. They said, "Have we any part in the affair?" Say you: "Indeed the affair belongs wholly to Allâh." They hide within themselves what they dare not reveal to you, saying: "If we had anything to do with the affair, none of us would have been killed here." Say: "Even if you had remained in your homes, those for whom death was decreed would certainly have gone forth to the place of their death," but that Allâh might test what is in your breasts; and to purifythat which was in your hearts (sins), and Allâh is All¬Knower of what is in (your) breasts. Those of you who turned back on the day the two hosts met (i.e. the battle of Uhud), it was Shaitân (Satan) who caused them to backslide (run away from the battlefield) because of some (sins) they had earned. But Allâh, indeed, has forgiven them. Surely, Allâh is Oft¬Forgiving, Most Forbearing. **O you who believe! Be not like those who disbelieve** (hypocrites) and who say to their brethren when they travel through the earth or go out to fight: "If they had stayed with us, they would not have died or been killed," so that Allâh may make it a cause of regret in their hearts. It is Allâh that gives life and causes death. And Allâh is All¬Seer of what you do. And if you are killed or die in the Way of Allâh, forgiveness and mercy from Allâh are far better than all that they amass (of worldly wealths). And whether you die, or are killed, verily, unto Allâh you shall be gathered."*

Contrary to the media announcement of the pagans Muhammad pbuh was not dead, yet the Muslim army suffered great losses. At the battle of Uhud 70-75 Muslims were killed in comparison to 22-37 pagans killed. The lesson Muslims learned from this was that even if Muhammad pbuh were to die and/or when he is dead, Muslims are not to abandon violent Jihad least of all when Muslims are getting killed fighting a defensive war in Muslim lands against the non-Muslims who are trying to stop Shariah and Islam from spreading. Allah told Muslims that flight from violence to the "safety" of unislamic non-Muslim lands was not the answer, but that Muslims were supposed to stand their ground and fight back defending whoever amongst the Muslims still remained even if worldly victory was not to be; that was their road to paradise. The apparent road to worldly safety was the path leading to disbelief and the eternal hellfire. After the battle of Uhud the pagan Meccan women cut off the ears and noses of the slain Muslims to make necklaces out of them. Muhammad's Uncle Hamza was among those slain and unfortunately Hamza's corpse was mutilated. Hamza's liver was chewed by the wife of Abu Sufyan and it is also reported that she ate his heart, his genitalia was also disfigured. Once the Muslims reorganized and learned Muhamad pbuh had not died but was just wounded they were furious and regretful over their poor performance in the battle and headed back to Medinah, where unknown to them the hypocrites had planned to fight them and prevent them from reentering the city. Before the Muslims returned however revelation was sent instructing Muhammad pbuh to go back to fight the pagan idolaters again. Despite their sorrow at the loss and inclination for revenge the Muslims were tired from battle fatigue, wounded and had lost 10% of their army, so they were not keen on fighting another battle. Yet because Allah ordered them to fight again without resting then they obeyed and tried to catch the pagan army before it returned to Mecca. To the surprise of the Muslims the pagan army met them on

the march. It turned out that the Meccan pagans figured
since Muhammad pbuh was still alive, as were many
zealous companions such as Abu Bakr and Umar then they
didn't really stop Islam or cripple the Muslim population
which was what their goal was, and since they had just won
a battle they thought it would be a prime time to attack the
Muslims in Medinah again. Although when they saw the
Muslims coming to them to fight the pagans realized the
Muslims were not bruised and beaten as they anticipated,
but eager for a rematch and it terrified the pagans. So the
pagans numbering over 2,963 fled from the 625 tired and
wounded Muslim army who they had defeated days before.
Thus while the Muslims lost the battle they won the field
and survived to worship Allah and practice Islam which was
their goal all along. Yet notwithstanding the safe return of
Muhammad pbuh and the Muslim army, all was not well
within the city of Medinah. After the battle of Badr the
pagans of Mecca wrote a letter to the Jewish tribe of Banu
Nadir living in Medinah, asking them to assassinate
Muhammad pbuh. While the Jews always had motive for
religious reasons, after the assassination of Ka'ab ibn Al-
Ishraf since he was a member of the Banu Nadir tribe, they
finally decided it was time to strike at Muhammad pbuh.
Two attempted assassination attempts the Jews of Banu
Nadir made on Muhammad pbuh are recorded in detail.
First the Jews proposed to Muhammad pbuh that he and 30
of the Muslims would meet 30 Jewish rabbis and they could
have a religious debate. The Jews said if the rabbis believed
what Muhammad pbuh said then all the Jews would
convert. Deciding to take the opportunity to preach Islam
despite knowing the Jews were insincere, in that they said "If

the rabbis believe then we will believe in Islam." instead of "We will believe in Islam if what you say is right.", Muhammad pbuh and 30 Muslims went to meet the Jews and 30 rabbis. Then the Jews asked Muhammad pbuh and 3 of his companions to come forth to the center of the public square to meet 3 rabbis. However before Muhammad pbuh came close to them a Muslim told Muhammad pbuh that his Jewish sister informed him the 3 rabbis had daggers and planned to kill him when he came close. Thus that plot of the Jews was foiled but it was hard to prove since the Rabbis could claim their daggers weren't for that purpose and the Jewish sister of the Muslim man had lied. Another report relates how the Jews of Banu Nadir requested that blood money be paid to them for 2 of their tribe who got accidentally killed by a Muslim after they had been ambushed with some being kidnapped outside of the city near Al-Raji. You see some people asked Muhammad pbuh to send them some Muslims to teach them Islam, so he did and then once the Muslims were far out of Medinah they abducted the Muslims and sent them to the Pagans of Mecca to be tortured and killed. Some of the Muslims in this party escaped and fought back, during which 2 Jews who were innocent of the kidnapping plot accidentally got killed for being in the wrong place at the wrong time. Therefore Muhammad pbuh expressed regret over the incident and went to handle the matter of paying blood money to the inheritors of the deceased. However the Banu Nadir did not want the blood money but were using this opportunity to kill Muhammad pbuh. While Muhammad pbuh sat next to a wall waiting for the Jews to meet with him, they were bringing a boulder to drop on him from atop the wall so the murder would seem accidental. But Allah informed Muhammad pbuh of this plot so that he was able to move away from the wall before any harm befell him. Of these 2 assassination attempts scholars differ over which took place first, personally I think the rock attempt happened before the dagger attempt because while the rock attempt is 100% confirmed as a plot due to revelation the rest of the Banu Nadir tribe could claim innocence of it. Thus in my opinion the plot wherein nearly all of Banu Nadir cooperated to have the Rabbis kill Muhammad pbuh seems to have been the latter plot that would more likely justify expulsion even though both plots in and of themselves violated the charter and merited expulsion. Regardless of which was first after the 2nd attempt, all agree that the Banu Nadir tribe were then besieged in their fortress. Al-Waqidi relates that the Jews of Banu Nadir were given a 10 day warning to leave the city but they refused and after the 10 days they settled into their fortress so the siege took place. Eventually the Banu Nadir surrendered and agreed to be expelled leaving their weapons behind in Medinah. Nearly everything else the Jews owned they took with them, including the very doors of their houses and fortress. Unlike with the Banu Qaynuqa the hypocrites did not side with the Jews against the Muslims. It was in regards to this event the 59th chapter of the Quran was revealed. Yet for brevity I will not quote it. However when the Banu Nadir left they went to Khaibar and were not content to leave the Muslims alone. Instead 50 leaders from the Jews of Banu Nadir, working with 20 Pagan leaders from Mecca signed a pact agreeing they would fight against Muhammad pbuh, Islam and Muslims until they died. Thereupon they gathered together a large coalition consisting of themselves and nearly everyone else in Arabia to launch an all-out assault on Medinah in the hopes of killing every Muslim and abolishing Islam. This battle took place in 5 A.H. or 627 CE and is referred to as the Battle of Ahzab. It was a 27 day long trench battle of attrition involving 3,000 Muslims vs. 10,000 disbelievers consisting of Pagans and Jews. The Jews who had been expelled from Medinah for trying to kill Muhamad pbuh were in this army, while the Jewish tribe of Banu Qurayzah still lived in Medinah. Duplicitously the Jewish tribe of Banu Qurayzah took this opportunity to betray the Muslims and colluded with their Jewish compatriots to attack the Muslim army from the rear and help the disbelievers surround the Muslim army trapping them between themselves and the trench that was blocking the army's march on Medinah. However this Jewish betrayal during the major battle of Muslims vs. Arabia did not succeed. After the Muslims sent a few hundred from the army back to Medinah to defend the city from the Jewish treason the Jews barricaded themselves in their fortresses. The Muslims later won the battle of Ahzab via a withdrawal by the coalition of disbelievers.

Then the siege of Banu Qurayzaah began and lasted for 25 days. Finally they agreed to surrender but they didn't want Muhammad pbuh to decide their fate but one of their own named Sa'd ibn Muad. He was a former Jew who had become Muslim and the Jews thought that because of his Jewish lineage then he would be biased towards his tribe and give them a much better deal than the expulsion they knew Muhammad pbuh would give them. They were wrong. Sa'd did not display favoritism for his former Jewish brethren but instead decide to judge them according to their own religious books(the Hebrew Bible and Talmud) and Jewish law wherein the penalty for actions such as they committed against the state was death. As a result of Sa'd's judgement 400-750 Jewish men of fighting ability were executed, the women were enslaved and the children who had yet to reach puberty were released. However 3 of these Jews became Muslim and 3 others were spared while one Jewish woman was executed for having killed a Muslim during the siege. After things calmed down a bit Muhammad pbuh and the Muslims decided to go to Mecca for a voluntary pilgrimage, so 1,400 of them went almost completely unarmed to Mecca. The Meccans didn't know what to do, because while they were at war with the Muslims the Muslims were peacefully trying to make a pilgrimage and the Meccans didn't want to just kill them because doing so would be dishonorable and destroy their tourism industry. However for the Pagans to let the Muslims into Mecca would cause them to lose honor too and may help to spread Islam. To resolve the situation a truce and treaty was formulated.

Muhammad pbuh then signed a treaty(via proxy since he couldn't read or write) with the Meccans agreeing to a truce for 10 years. This was called the treaty of Hudahbiyyah and was signed in 6 A.H. or 628 CE. Afterwhich in 628 CE, the Muslims conquered the Jews who settled in Khaibar, Arabia because they had joined and gathered the army that previously fought the Muslims at the battle of Ahzab and those Jews didn't sign a peace treaty as the pagans did. The conquest of Khaibar resulted in 93 Jewish warriors dead with 15-20 Muslim warriors killed. Then the pagans violently broke the truce in 8 A.H. or 629 CE when they realized Islam was peacefully spreading at an exponential rate. As is evident in the numbers of Muslim soldiers swelling from 314 to 3,000 in just 4 years, the global Muslim population was increasing at a rate over 25% per year. Meaning for every 4 Muslims that existed, by the end of each year there would be 5. Throughout this period every day people saw the moral behavior of Muslims and heard the pure undiluted and undistorted message of Islam. The more Islam was being practiced more and more people were voluntarily embracing Islam. Even amongst pagans many were becoming Muslims after acknowledging it's truth. Some of the staunchest opponents of Islam ended up becoming some of the best leaders of the Muslims. Khaalid Ibn Waleed is one example. 4 years after Khaalid defeated and ruthlessly butchered the Muslims during the Battle of Uhud this same general came on his own from Mecca to Medinah to testify his belief that, *"There is nothing to be worshiped except God and Muhammad is the Messenger of God"*. Khaalid was afraid the Muslims might have sore feelings towards him because of his past slaughtering of them, but all was forgiven. Today Khaalid Ibn Waleed is known as a prolific Muslim who is highly esteemed throughout the Muslim world, despite having fought against the Muslims and the prophet pbuh in battles before he became a Muslim. Now some may be thinking if Khaalid was a military genius who never lost a battle then maybe the Muslim conquests were because of him and had nothing to do with Allah. Unfortunately some Muslims even mistakenly think this and believe military might or technology is the cause for the Muslim's position today. However anyone who thinks Khaalid was responsible for victory is in deep error. This is because 1. Victory only comes from Allah. 2. Islam says that the sins of Muslims is the only reason they can ever lose and that there will usually always be more disbelievers and they will usually always have better weapons when fighting the Muslims, so that the only way Allah will give the Muslims victory is if they refrain from sins. Therefore the very first thing Umar bin Khattab did as Khalifah was to fire Khaalid from his position as general and demote him to being a regular soldier. Khaalid only became a Muslim on May 31, 629 CE, he became a general of the Muslims during the Battle of Mut'ah in September 629 CE after the top 3 generals of the Muslims were killed in battle with the Christian Byzantine Empire and on August 22, 634 CE Umar dismissed Khaalid from his position of general. So Khaalid was only a general of the Muslims for less than 5 years, it was a very active and illustrious 5 years but he was not the reason for the Muslim conquests despite being a top-notch general. In 634 CE Khaalid was no longer a general. The reason that Umar demoted Khaalid was because he was afraid Muslims would think victory came due to Khaalid and not because of Allah, thus Umar fired Khaalid because he was too good at war. This is what Umar actually said, he told Khaalid that because he never loses then it's religiously dangerous for him to be a general since it could lead to hero worship, thus in order to keep Islamic theology pure from extremism or corruption it's best general was dismissed in the prime of his career. Now does that sound like the actions of a religion which spreads it's faith via the sword?

Later the Muslims proved through further conquests that victory only comes from Allah and has nothing to do with military numbers or technology. Now I purposely didn't say strategy because strategy does determine the outcome, the winning strategy is: *"If you believe in Islam and don't sin, then when it's time to do battle Allah will make you win."* By this time, 629 CE, Muslims were no longer a small minority as they were when Muhammad pbuh left Mecca in 622 CE. The pagan Meccans feared the growing strength of the Muslims and the peaceful spread of Islam so they felt they had no choice but to break the treaty and wage war lest Islam spread even further. Then verses of the Quran were revealed instructing the Muslims what to do in response to the pagans betrayal and violation of the peace treaty. Some of the verses revealed instructing Muslims what to do when the Meccans broke the treaty are in Chapter 9 verses 1-33, I've included verses revealed later about Jews and Christians as well to give both a historical and modern context of the Quran.

Freedom from (all) obligations (is declared) from Allâh and His Messenger to those of the Mushrikûn (polytheists, pagans, idolaters, disbelievers in the Oneness of Allâh), with whom you made a treaty. So travel freely (O Mushrikûn) for four months (as you will) throughout the land, but know that you cannot escape (from the Punishment of) Allâh, and Allâh will disgrace the disbelievers. And a declaration from Allâh and His Messenger to mankind on the greatest day (the 10th of Dhul-Hijjah - the 12th month of Islâmic calendar) that Allâh is free from (all) obligations to the Mushrikûn and so is His Messenger. So if you (Mushrikûn) repent, it is better for you, but if you turn away, then know that you cannot escape (from the Punishment of) Allâh. And give tidings (O Muhammad) of a painful torment to those who disbelieve. Except those of the Mushrikûn with whom you have a treaty, and who have not subsequently failed you in aught, nor have supported anyone against you. So fulfill their treaty to them for the end of their term. Surely Allâh loves Al- Mattaqûn (the pious). Then when the Sacred Months (the 1st, 7th, 11th, and 12th months of the Islâmic calendar) have passed, then kill the Mushrikûn wherever you find them, and capture them and besiege them, and lie in wait for them in each and every ambush. But if they repent and perform As-Salât (Prayer), and give Zakât, then leave their way free. Verily, Allâh is Oft-Forgiving, Most Merciful. And if anyone of the Mushrikûn (polytheists, idolaters, pagans, disbelievers in the Oneness of Allâh) seeks your protection then grant him protection, so that he may hear the Word of Allâh (the Qur'ân), and then escort him to where he can be secure, that is because they are men who know not. How can there be a covenant with Allâh and with His Messenger for the Mushrikûn (polytheists, idolaters, pagans, disbelievers in the Oneness of Allâh) except those with whom you made a covenant near Al-Masjid-al-Harâm (at Makkah)? So long, as they are true to you, stand you true to them. Verily, Allâh loves Al-Muttaqûn (the pious) How (can there be such a covenant with them) that when you are overpowered by them, they regard not the ties, either of kinship or of covenant with you? With (good words from) their mouths they please you, but their hearts are averse to you, and most of them are Fâsiqûn (rebellious, disobedient to Allâh). They have purchased with the Ayât (proofs, evidences, verses, lessons, signs, revelations, etc.) of Allâh a little gain, and they hindered men from His Way; evil indeed is that which they used to do. With regard to a believer, they respect not the ties, either of kinship or of covenant! It is they who are the transgressors. But if they repent, perform As-Salât (Prayer) and give Zakât, then they are

*your brethren in religion. (In this way) We explain the Ayât
(proofs, evidences, verses, lessons, signs, revelations, etc.) in detail
for a people who know. But if they violate their oaths after their
covenant, and attack your religion with disapproval and criticism
then fight (you) the leaders of disbelief (chiefs of Quraish - pagans
of Makkah) - for surely their oaths are nothing to them - so that
they may stop (evil actions). Will you not fight a people who have
violated their oaths (pagans of Makkah) and intended to expel the
Messenger, while they did attack you first? Do you fear them?
Allâh has more right that you should fear Him, if you are believers.
Fight against them so that Allâh will punish them by your hands
and disgrace them and give you victory over them and heal the
breasts of a believing people, And remove the anger of their
(believers') hearts. Allâh accepts the repentance of whom He wills.
Allâh is All-Knowing, All-Wise. Do you think that you shall be
left alone while Allâh has not yet tested those among you who have
striven hard and fought and have not taken Walîjah [(Batanah -
helpers, advisors and consultants from disbelievers, pagans) giving
openly to them their secrets] besides Allâh and His Messenger, and
the believers. Allâh is Well-Acquainted with what you do. It is not
for the Mushrikûn (polytheists, idolaters, pagans, disbelievers in
the Oneness of Allâh), to maintain the Masjids of Allâh (i.e. to
pray and worship Allâh therein, to look after their cleanliness and
their building), while they witness against their ownselves of
disbelief. The works of such are in vain and in Fire shall they abide.
The Masjids of Allâh shall be maintained only by those who believe
in Allâh and the Last Day; perform As-Salât (Prayer), and give
Zakât and fear none but Allâh. It is they who are on true guidance.
Do you consider the providing of drinking water to the pilgrims
and the maintenance of Al-Masjid-al-Harâm (at Makkah) as equal
to the worth of those who believe in Allâh and the Last Day, and
strive hard and fight in the Cause of Allâh? They are not equal*

before Allâh. And Allâh guides not those people who are the
Zâlimûn (polytheists and wrong-doers). Those who believed (in
the Oneness of Allâh - Islâmic Monotheism) and emigrated and
strove hard and fought in Allâh's Cause with their wealth and
their lives are far higher in degree with Allâh. They are the
successful. Their Lord gives them glad tidings of Mercy from Him,
and that His being pleased (with them), and of Gardens (Paradise)
for them wherein are everlasting delights. They will dwell therein
forever. Verily, with Allâh is a great reward. O you who believe!
Take not for Auliyâ' (friends, supporters and helpers) your fathers
and your brothers if they prefer disbelief to Belief. And whoever of
you does so, then he is one of the Zâlimûn (wrong-doers). Say: If
your fathers, your sons, your brothers, your wives, your kindred,
the wealth that you have gained, the commerce in which you fear a
decline, and the dwellings in which you delight are dearer to you
than Allâh and His Messenger, and striving hard and fighting in
His Cause, then wait until Allâh brings about His Decision
(torment). And Allâh guides not the people who are Al-Fâsiqûn
(the rebellious, disobedient to Allâh) Truly Allâh has given you
victory on many battlefields, and on the Day of Hunain (battle)
when you rejoiced at your great number but it availed you naught
and the earth, vast as it is, was straitened for you, then you turned
back in flight. Then Allâh did send down His Sakînah (calmness,
tranquillity and reassurance, etc.) on the Messenger
(Muhammad), and on the believers, and sent down forces (angels)
which you saw not, and punished the disbelievers. Such is the
recompense of disbelievers. Then after that Allâh will accept the
repentance of whom He wills. And Allâh is Oft-Forgiving, Most
Merciful. O you who believe (in Allâh's Oneness and in His
Messenger (Muhammad)! Verily, the Mushrikûn (polytheists,
pagans, idolaters, disbelievers in the Oneness of Allâh, and in the
Message of Muhammad) are Najasun (impure). So let them not

come near Al-Masjid-al-Harâm (at Makkah) after this year, and if you fear poverty, Allâh will enrich you if He wills, out of His Bounty. Surely, Allâh is All-Knowing, All-Wise. Fight against those who (1) believe not in Allâh, (2) nor in the Last Day, (3) nor forbid that which has been forbidden by Allâh and His Messenger (Muhammad) (4) and those who acknowledge not the religion of truth (i.e. Islâm) among the people of the Scripture (Jews and Christians), until they pay the Jizyah with willing submission, and feel themselves subdued. And the Jews say: 'Uzair (Ezra) is the son of Allâh, and the Christians say: the Messiah is the son of Allâh. That is their saying with their mouths, resembling the saying of the those who disbelieved aforetime. Allâh's Curse be on them, how they are deluded away from the truth! They (Jews and Christians) took their rabbis and their monks to be their lords besides Allâh (by obeying them in things which they made lawful or unlawful according to their own desires without being ordered by Allâh), and (they also took as their Lord) Messiah, son of Maryam (Mary), while they (Jews and Christians) were commanded [in the Taurât and the Injeel to worship none but One Ilâh (God - Allâh) Lâ ilâha illa Huwa (none has the right to be worshipped but He). Praise and glory is to Him, (far above is He) from having the partners they associate (with Him)." They (the disbelievers, the Jews and the Christians) want to extinguish Allâh's Light (with which Muhammad has been sent - Islâmic Monotheism) with their mouths, but Allâh will not allow except that His Light should be perfected even though the Kâfirûn (disbelievers) hate (it). It is He Who has sent His Messenger (Muhammad) with guidance and the religion of truth (Islâm), to make it superior over all religions even though the Mushrikûn (polytheists, pagans, idolaters, disbelievers) hate (it)."

After the pagans broke the truce the Muslims gave them a warning in advance that they were at war again and hostilities would commence. Technically speaking the Muslims were always at spiritual war with the pagans but they had not been at physical war due to the truce. So the declaration of war was dealing with the physical militaristic aspect of war, rather than the spiritual war in which there is never any truce or cessation. The Muslim army numbering 10,000 marched to Mecca, with it being the first time back in Mecca for some Muslims, returning after they had been forced to leave many years earlier. Remember 45 Muslims from Mecca migrated to Medinah in 622 CE and the combined count of Muslim men in 622 CE at Madinah was 90. Just 7 years later in 629 CE those Muslim refugees came back to the pagan Meccans with an army of 10,000 Muslims. It was an event they never imagined possible and it terrified them. To put it in modern perspective it is alleged there are 3.3 million Muslims in America, imagine if they all left because of violent persecution and then came back 7 years later with an army of 366.3 million. Since in total there are only about 330 million Americans it's easy to imagine the non-Muslim Americans would be terrorized in such a scenario. Honestly they'd call it the Apocalypse. Likewise the Muslim army marching on Mecca led by Muhammad pbuh had more soldiers than there were people in Mecca, so the result was easy to pre-determine. The Muslim troops were given explicit instructions by the Prophet pbuh to only raise their hands against those who obstructed their entrance into Mecca or attacked them. They were ordered not to touch any moveable or immovable property, or destroy anything of the Meccans; in accordance to the rules of jihad.

There were also 10 names Muhammad pbuh gave of individuals (6 men and 4 women) who would be killed on sight wherever they were because they were guilty of committing war crimes. The Muslim army surrounded Mecca from all sides prior to entering. Most of the Muslim army entered in peace, but one part of the city resisted and 2 Muslims were killed and 12 of the pagan aggressors, but it was relatively a peaceful surrender of Mecca without bloodshed; aside from the few who resisted and some war criminals who were put to death for past murders committed. Of the 10 war criminals only 3 men and 2 women were actually killed, the rest were pardoned because they had embraced Islam. During the conquest of Mecca 2 Muslim soldiers and 17 disbelievers were killed, making 19 total casualties. Riding on a camel into the city, Muhammad pbuh was not boastful or arrogant. Rather he was lowering his head with so much humility that his forehead was nearly touching the saddle, while he recited verses of the Quran which attributed victory to the will of Allah. At the Kaba, which had been built by Abraham pbuh who started the annual pilgrimage of Hajj, the Meccans had placed many idols inside and around the sacred house having corrupted the place to entice pagans to spend money coming on a pilgrimage to Mecca. With Mecca now in Muslim control, one by one the idols were destroyed until the Kaba and Mecca was liberated from idolatry. Muhammad pbuh then asked for all the people of Mecca to be assembled before him, and they came. He then proceeded to calmly remind them of all the abuses and crimes they were guilty of perpetrating against Allah, Islam, himself and Muslims for the last 21 years. He asked "Do you remember doing this?"

and "Do you remember doing that?", "Do you remember saying X about me?" and they confessed everything since they knew it was all true. The pagans of Mecca then anxiously asked what was to be done to them, now that they had been conquered by those very same Muslims whom they treated so harshly who seemed rather pissed off and now were the conquerors. They rightly feared punishment for their previous oppression, just as the enemies of Islam today are terrorized by the thought of what will happen when Allah puts the Muslims in power over them. In response to them Muhammad pbuh replied, *"I speak to you in the same words of the prophet Yusuf (Joseph) peace be with him, who spoke to his brothers"*. Then Muhammad pbuh recited a verse from the Quran relaying what Yusuf (Joseph) pbuh said, which means: *"no blame will there be upon you today. Allah will forgive you; and he is the most merciful of the merciful"*. These pagan meccans had slandered him, killed his family, friends and those who embraced his message of islam, boycotted muslims, confiscated their property, attacked him and his followers on a daily basis for years and even chased Muhammad pbuh out of town! Muhammad remembered it all, they even admitted they were guilty and didn't even pretend to apologize for it. Had the inhabitants of mecca been slaughtered in a similar fashion as to how the crusaders butchered the people when they conquered Jerusalem, Muhammad pbuh would have been considered fair and justified in doing so, historically and in his own time. But Islam is not a religion of revenge, it is a religion of submission to the one almighty creator of all things. For a Muslim victory is: worshipping God, living as God wants, establishing Islam and entering paradise. All that a Muslim does should be done with the intention of pleasing God. In such a situation forgiveness is more pleasing than extracting justice and many of the meccans voluntarily embraced Islam as a result of this mercy which only a prophet would be capable of. To the pagan meccans who rejected nearly every sign of Muhammad's prophethood and numerous miracles, his mercy was the miracle they accepted that convinced them he really was a prophet of god and Islam was the only true religion. However not everyone accepted Islam and those who didn't weren't harmed, which is a testament that there was no compulsion and that all who embraced Islam really did so with sincerity. Since Mecca is a blessed city unlike any other, the polytheistic pagans were not allowed to live there due to Mecca's special status. Thus the polytheists were told they had 3 months to decide whether they wanted to become Muslims or not and if not then they would have to move somewhere else because Mecca was a special city where polytheists are not permitted. It was not a convert or die scenario, except for the 10 war criminals who legally earned a death penalty by all accounts half of whom chose to embrace Islam. Yet that rule of moving is not the standard rule, that was for Mecca because it's Mecca. Christians and jews can peacefully remain in a Muslim controlled city(except Mecca) and pay jizya without having to move or change their religion. At the end of it all conquest is a fact of life, we can't pretend that people will not fight and conquer each other. You might wish that were the case but reality is reality. Wars will happen and peoples will get conquered, the important thing is how does God desire us to act when that happens? When x people conquers y people for the sake of their differences in religion then how are they supposed to act when they win? Is there any possible way to improve even a little bit on how Muhammad pbuh taught us to act? I don't believe so. Does Judaism or Christianity or any other ideology teach us how to act in a way that is truly just from all perspectives? No. There is not another group in history who can claim they waged war in a way which God was pleased with and everyone can agree they are correct to say so. No nation except for the Muslim nation can say they fought and conquered people according to how God desired them to. Certainly the modern nations today can't claim their wars are sanctioned by god, either before, during or after. Why? What makes the Muslims different? Because they submit to god as slaves. A true slave of God does not do anything at all unless it is in conformity with God's desires and rules. If God doesn't like it then a slave of God doesn't do it. The reason why opponents of Islam want to distort its image and make people have misconceptions about it is because of this submission to the Creator and liberation from everything else. Islam frees people from idolatry, vices, addiction, sin, over-indulgence and many other things that are bad for society but lucrative for special interests. Regardless of what one believes, they are a slave. Either to their country, their desires, their money, their peers, their fears or other things. Islam frees people from all of that and more. Islam frees people from corrupt man-made systems and oppressive false religions, establishing justice and morality throughout the world for everyone. Islam frees people from the flaws and false promises of "freedom". Islam frees people from racism, tribalism, exploitation, nationalism, foolish family customs and traditions, social conformity, sadism, misogynism, ethnocentrism, consumerism and every wicked thing that humanity has ever come across or come up with, or ever will. But most importantly Islam can free the individual who sincerely practices it and dies upon it from the hellfire and cause them to enter paradise. This is why satan and all his forces always have and always will incite intolerance towards Islam & Muslims, because they cannot tolerate humans correctly worshipping and obeying solely the one who made them being rewarded for it for eternity. Whereas Muslims will never tolerate oppression, falsehood, corruption or evil of any kind at any time in any place. So Jihad will continue as long as Satan wages war against mankind and as long as mankind wages war against itself.

Narrated Abdullah ibn Umar:

I heard the Messenger of Allah, say: When you enter into the inah transaction, hold the tails of oxen, are pleased with agriculture, and give up conducting jihad (struggle in the way of Allah). Allah will make disgrace prevail over you, and will not withdraw it until you return to your original religion.

Source: Sunan Abi Dawud 3462 Graded Sahih by Albani

Narrated Abu Umamah:

The Prophet said: He who does not join the warlike expedition (jihad), or equip, or looks well after a warrior's family when he is away, will be smitten by Allah with a sudden calamity. Yazid ibn Abdu Rabbihi said in his tradition: 'before the Day of Resurrection".

Source: Sunan Abi Dawud 2503 Graded Hasan by Albani

www.ingramcontent.com/pod-product-compliance
Lightning Source LLC
Chambersburg PA
CBHW071758120626
46550CB00002B/838